OXFORD HISTORICAL MONOGRAPHS

THE DUKERIES TRANSFORMED

The Social and Political Development of a Twentieth Century Coalfield

Robert J. Waller

CLARENDON PRESS · OXFORD
1983

Oxford University Press, Walton Street, Oxford OX2 6DP
London Glasgow New York Toronto
Delhi Bombay Calcutta Madras Karachi
Kuala Lumpur Singapore Hong Kong Tokyo
Nairobi Dar es Salaam Cape Town
Melbourne Auckland
and associated companies in
Beirut Berlin Ibadan Mexico City Nicosia

Oxford is a trade mark of Oxford University Press

Published in the United States
by Oxford University Press, New York

British Library Cataloguing in Publication Data
Waller, Robert J.
The Dukeries transformed: the social and political development of a twentieth century coalfield. — (Oxford historical monographs)
1. Coal trade — England — Sherwood Forest region (Nottinghamshire)
I. Title
338.2'724'0942521 HD9551.7.N/
ISBN 0-19-821896-6

Library of Congress Cataloging in Publication Data
Waller, Robert J.
The Dukeries transformed; the social and political development of a twentieth-century coalfield.
(Oxford historical monographs)
Bibliography: p.
Includes index.
1. Coal-miners — England Nottinghamshire — History — 20th century. 2. Coal mines and mining — England — Nottinghamshire — History — 20th century. 3. Nottinghamshire — Economic conditions. 4. Nottinghamshire — Social conditions. I. Title.
HD8039.M62G78 1983 338.2'724'094252 83-4195
ISBN 0-19-821896-6

Typeset by Hope Services, Abingdon
Printed in Great Britain
at the University Press, Oxford
by Eric Buckley
Printer to the University

To
MMW

Acknowledgements

I would like to thank the Social Sciences Research Council, and Merton and Magdalen Colleges for funding my work between 1977 and 1982. I am most grateful to my supervisor, Pat Thompson, to all the interviewees, and to so many who have been of such help, especially the following: Paul E. Bailey, Ernest Baker, Stanley Chapman, Ted Corke, Ewan Ferlie, Alan Griffin, Colin Griffin, Brian Harrison, David Howell, Patrick Joyce, Jim Lodge and British Steel Stanton, Steven Lukes, Ross McKibbin, Howard Newby, Alan Rogers, Barry Supple, Paul Turner, Philip Waller, and Stephen Wright.

Contents

Figures

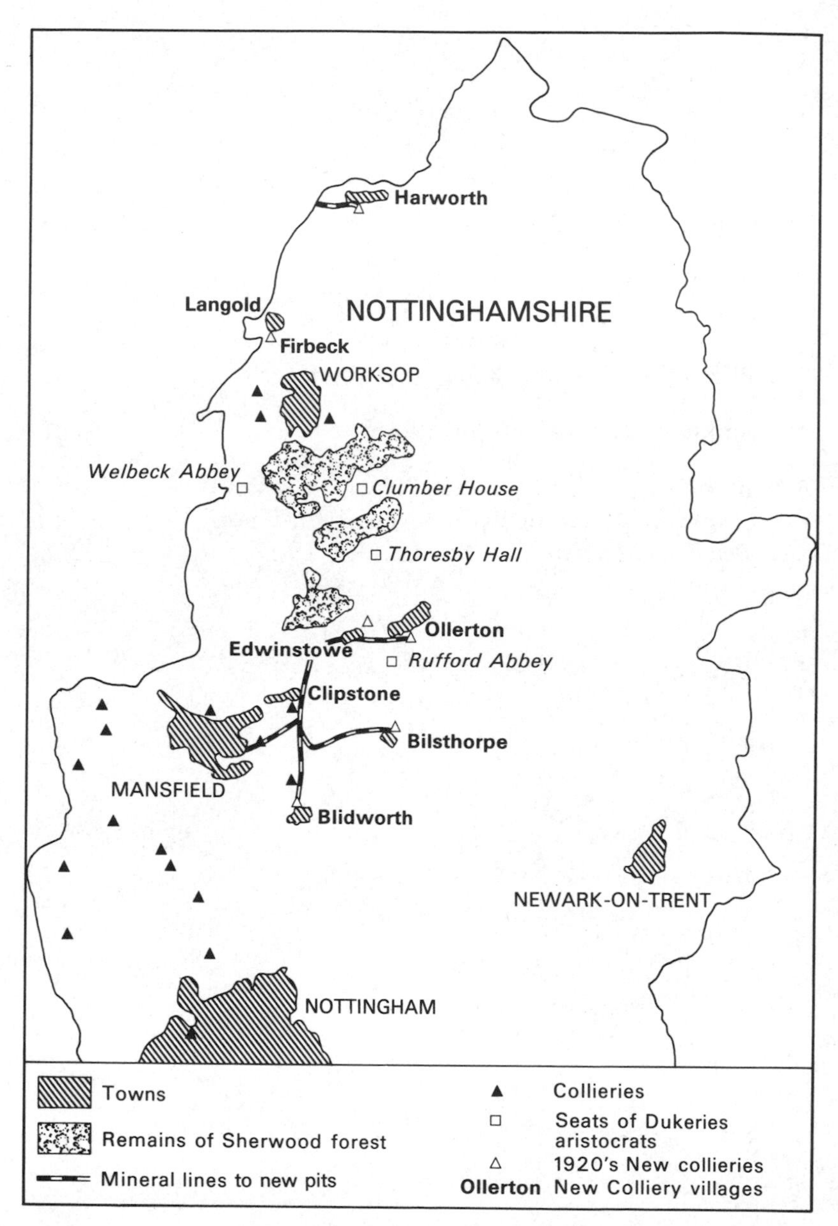

Fig. 1 The Dukeries

Introduction

Far more attention has been paid to the social and economic effects of pit closure in the British coalfields than to the effects of the opening of new pits and the development of new fields. The reason for this is not hard to understand. As far as the twentieth century is concerned, the period for which we have most evidence and which might be considered the most relevant to the future course of the coal industry, the received picture of the coalfields is either one of a stable stereotype, or more likely one of stagnation and decline. If anything is seen as disturbing the pattern of social homogeneity and cohesion allied to a high level of political and industrial organization,[1] it is the impact of the closing of mines and the consequent changes in the character of the region concerned. Yet the conclusions of such studies as those by the Newcastle University group of J. W. House and E. M. Knight,[2] and by Martin Bulmer in the case of Spennymoor,[3] suggest that the dislocation of places undergoing pit closure, though great, may not be crippling. The distinctive nature of a mining village may not be eradicated by the removal of its traditional economic base. In this book it is intended to be shown that the impact of the *opening* of pits in previously undeveloped coalfields certainly can cause a very high degree of social and political dislocation, besides the more obvious effects of the economic transformation of the locality.

It is, of course, true that the overall pattern of the British coal industry in the twentieth century has been one of

[1] For a sociological account of this pattern, see N. Dennis, E. Henriques and C. Slaughter, *Coal is Our Life* (London, 1956).

[2] J. W. House and E. M. Knight, 'Pit Closure and the Community', *Papers on Migration and Mobility in Northern England*, no. 5, Dec. 1967; E. M. Knight, 'Men Leaving Mining, West Cumberland', *Papers on Migration and Mobility in Northern England* no. 6, Jan. 1968. See also *Ryhope, a Pit Closes* (HMSO, 1970).

[3] Martin Bulmer (ed.), *Mining and Social Change: Durham County in the 20th Century* (London, 1978), ch. 14.

decline. Production and employment figures for the whole country show a steady decline from their peaks in the years just before and after the First World War. But concealed in these statistics are considerable developments in individual coalfields. In the inter-war years the best known of these was probably the new field in Kent, but a more extensive expansion took place in the Dukeries district of Nottinghamshire where the eastern extension of the main Midlands field was exploited for the first time in the 1920s and 1930s.[4] Tables 1

Table 1. *Production of coal in various coalfields 1913-46*[a]
(tons × 1,000)

	Notts.	Durham	S. Wales	Lancs.	Kent	N. Derby.	U.K.
1913	12,394	41,533	56,830	24,629	59	16,859	287,430
1923	14,217	38,218	54,252	20,155	488	16,065	276,001
1924	14,190	36,690	51,085	19,790	330	15,603	267,118
1925	14,022	31,493	49,630	17,422	368	14,724	243,176
1926	9,088	14,136	20,273	9,170	214	9,164	126,279
1927	13,647	34,603	46,256	17,106	637	13,745	251,232
1928	13,194	34,709	43,312	15,076	930	13,093	237,472
1929	14,738	39,001	48,150	15,659	1,149	14,199	256,907
1930	14,475	35,863	45,108	15,004	1,292	13,788	243,882
1931	14,483	30,249	37,085	14,115	1,586	12,646	219,459
1932	13,550	27,802	34,874	13,247	1,824	11,837	208,753
1933	13,638	27,606	34,355	13,205	1,928	11,239	207,112
1934	14,309	30,590	35,173	13,758	2,030	11,798	220,727
1935	14,015	30,273	35,025	14,146	2,089	11,979	222,253
1936	15,059	31,372	33,886	14,660	2,026	12,471	228,453
1937			35,309				233,192
1938	15,468	32,668	35,293	14,285	1,771	13,013	226,993
1939	16,763	30,649	35,269	14,320	1,865	13,805	231,338
1940	17,673	26,709	32,352	13,617	1,572	14,854	224,299
1941	17,829	24,599	27,426	12,245	1,377	14,624	206,344
1942	17,664	25,339	26,723	11,700	1,322	14,802	203,633
1943	16,479	24,347	25,116	11,146	1,388	14,469	194,443
1944	15,273	22,767	21,569	10,708	1,313	13,071	181,410
1945	15,303	21,678	19,482	10,353	1,213	13,027	171,923
1946	14,576	22,514	19,088	10,684	1,301	12,525	175,712

[a] Parliamentary Papers, 1937-8 xxvii, 330-1; PP 1943-4 viii, 163-4; PP 1945-6 xxi, 17.

[4] See below, ch. 1.

and 2 give the first hint of a break in the typical pattern of the industry in these years, although it must be remembered that the figures for the county of Nottinghamshire are influenced by the decline of the old coalfield on the Derbyshire border. Despite this handicap, the county as a whole shot up the 'league tables' of employment and production, showing an ability to hold its own or improve its position while other fields were in rapid decline. This improvement was almost entirely due to the growth of the new eastern Sherwood Forest field where seven new mines were sunk between 1918 and 1928, as listed in Table 3.

Table 2. *Employment in selected coalfields 1913–43*[a]
(men employed × 1,000)

	Notts.	Durham	S. Wales	Lancs.	Kent	N. Derby.	U.K.
1913	40.5	165.2	232.8	107.7	1.1	55.4	1105.2
1923	55.4	174.1	252.6	106.4	2.1	61.0	1203.4
1924	57.4	174.8	250.1	105.6	1.7	62.4	1213.9
1925	57.2	143.0	217.8	99.3	1.9	60.1	1102.6
1926	58.5	147.2	217.8	98.9	2.1	59.2	1115.8
1927	58.0	130.7	194.1	90.6	2.8	56.9	1023.9
1928	52.1	130.2	168.3	81.3	3.6	53.1	939.0
1929	52.7	133.8	178.3	79.2	4.3	53.0	956.7
1930	52.4	133.3	172.9	75.7	5.1	52.7	931.3
1931	51.3	115.2	158.2	72.5	5.7	50.5	867.9
1932	49.5	106.0	145.7	67.8	6.4	48.3	819.3
1933	47.0	102.9	142.9	65.4	6.6	44.6	789.1
1934	46.9	107.9	139.8	62.3	7.1	43.9	788.2
1935	45.9	107.1	131.7	60.7	7.3	42.7	769.5
1936	45.5	108.7	126.2	60.1	7.4	41.7	767.1
1937	46.4	109.3	124.6	60.2	7.3	42.8	739.2
1938	45.1	114.5	134.8	58.3	6.6	41.6	781.7
1939	45.6	111.2	128.8	56.3	6.4	41.4	766.3
1940	45.5	105.4	128.5	52.6	5.7	41.0	749.2
1941	44.6	93.0	111.6	48.5	5.0	40.2	697.6
1942	44.8	98.8	114.2	49.0	5.3	40.4	709.3
1943	44.2	100.0	114.3	49.6	5.5	40.2	707.8

[a] PP 1937–8 xxvii, pp. 332–3; PP 1943–4 viii, pp. 163–4; PP 1945–6 p. 17.

Table 3. *New pits, Nottinghamshire 1918–39*

	sinking commenced	coal won
Clipstone	1920	1922
Harworth	1919	1924
Firbeck Main	1922	1925
Ollerton	1923	1925
Blidworth	1924	1926
Bilsthorpe	1925	1927
Thoresby	1925	1928

The vast resources detected beneath the remnants of the forest led to high hopes among the coal companies that the new pits would bring great profits for many decades. Although the most extravagant claims for the new field were not fulfilled, the Dukeries did become the most productive and profitable coalfield in the country, with the possible exception of South Yorkshire. It has remained a mainstay of the nationalized industry while traditional centres of mining such as Durham and South Wales have dwindled in importance. By 1980 more coal was being produced in Nottinghamshire than in the North-East of England and South Wales put together.[5]

The deep pits sunk in eastern Nottinghamshire in the 1920s proved to be larger and more productive than older, smaller mines. In the North Midlands Survey Report of 1945, it was discovered that in Nottinghamshire and Derbyshire, the sixty-six collieries then in operation in the western exposed field, where the coal measures outcrop on the surface, employed approximately 40,000 men at an average of 603 per pit. The twenty-six collieries of the older concealed coalfield on the Notts.–Derby. border employed 28,000 at an average of 1,070 per pit. The eight collieries of the new concealed coalfield studied in this book employed over 12,000 men at an average of 1,537 each.[6] The profitability

[5] NCB Annual Report 1980–1, pp. 34–5.

[6] *North Midland Regional Survey Report* (HMSO, 1945), p. 22.

of the Dukeries pits has enabled the National Coal Board to retain in service many of the smaller, older, loss-making pits. In 1977, for example, Thoresby Colliery produced 1,269,000 tons of coal, whereas Sutton Colliery, sunk in 1874, produced only 362,000 tons, and Silverhill (1875) 465,000 tons.[7]

Of course, not only were new pits sunk when a field such as this was developed. The provision of housing for the miners to be employed was regarded as an essential part of their total investment by the coal companies involved, who reckoned to reserve as much as 45 per cent of their capital expenditure for building new colliery villages 'at the pit gates' to ensure a guaranteed supply of steady labour.[8] As Table 4 shows, this resulted in the mushroom growth of more than half a dozen communites in the Dukeries, some attached to old agricultural villages, some completely new. The nature and problems of these new villages will form the principal subjects of this book.

Table 4. *Population of mining parishes, Nottinghamshire 1921–31*[a]

	1921	1931
Southwell RD	(20,157)	(32,076)
Bilsthorpe	134	1,972
Clipstone	592	3,443
Boughton	315	1,318
Edwinstowe	963	2,818
Ollerton	676	3,912
Worksop RD	(5,070)	(14,555)
Harworth	865	6,092
Hodsock	232	4,307
Skegby RD		
Blidworth	2,033	5,316

[a] Source: *1931 Census, Nottinghamshire*, Table 2, pp. 6–7.

[7] National Coal Board, *Coal in North Nottinghamshire*, 1977–8.

[8] P. H. White, 'Some Aspects of Urban Development by Colliery Companies 1919–39', *Manchester School of Economic and Social Studies*, 23 (1955), 269–80.

It must be remembered that the exploitation of the new Nottinghamshire field involved a considerable migration of labour. It is necessary to enquire into the origins of the immigrant miners, their reactions to their new environment, and the development of a community spirit in the 'cosmopolitan' and rapidly growing colliery villages.[9] One of the arguments central to Martin Bulmer's attempt at a sociological model of a mining community[10] states that little geographical mobility is to be found in coal-mining districts. Here, as elsewhere, it may be suggested that pit opening produces conditions which conflict with Bulmer's ideal typification. A high turnover of population in the new coalfield does also imply that the dislocations apparent in the early history of the Dukeries colliery villages affected far more people than at first appears.

The development of the new coalfield in the Dukeries may also be considered as an example of the industrialization of a previously rural area where the economy had been dominated by agriculture and by the great estates of the aristocrats who gave the district its popular name: the Dukes of Newcastle and Portland, Earl Manvers, Viscount Galway, and Lord Savile. It is of interest to study contemporary reactions to the arrival of coal mining in the Dukeries, and particularly how the existing inhabitants of the neighbourhood regarded the great changes which were taking place. What were relations like between the residents of the newly established colliery villages and those of the old villages? What attitude did the Dukeries landowners adopt, when faced by a transformation which might considerably alter the aspect of their country seats, and yet at the same time bring them much-needed mineral royalties?

Another central theme of the study will be the nature of the colliery villages, purpose built at the pit gates to house the employees of the coal-owners. To what extent can they be characterized as company towns, in which the influence of the employers extended beyond the hours of work at the pit itself to the social and political life of the village? The homogeneous quality of these isolated villages, which could

[9] See below, ch. 2.

[10] M. I. A. Bulmer, 'Sociological Models of the Mining Community', *Sociological Review*, NS 23 (1975), 61-92.

offer no alternative source of employment or housing to that provided by the company, gave the owners a potential for control of the local community increasingly unusual in twentieth-century Britain. The general economic conditions of the 1930s also gave added weight to the ultimate weapon of the employer, the threat of dismissal. The effect of these factors on industrial relations and trade unionism in the new Nottinghamshire coalfield is especially interesting because of the existence in that county of George Spencer's 'non-political' trade union, set up in opposition to the official Nottinghamshire Miners Association after the collapse of the coal strike of 1926. The new coalfield became a stronghold of Spencerism, and not a single strike was called before the Second World War. Similarly, political militancy was hindered by the miners' weak social and economic position. The Labour party was not established in the colliery villages before the war, and management allied itself with the old village and county élites to maintain Conservative representation and control for twenty years after the arrival of mining in the Dukeries.

In other ways too the early years of the Dukeries coalfield were marked by disruption and dislocation in the lives of its inhabitants. The local authorities had to deal with the difficulties of supplying essential services such as roads, sewerage systems, water, and schools for the new villages. Here the clashes of authority and the teething troubles consequent upon the foundation of any new community can be seen clearly, exacerbated in this case by the fact that the Rural District Councils responsible were dominated by councillors from rural parishes, and neither elected representatives nor officers of the Council had any experience of planning the major schemes necessary to cater for a burgeoning population. The effects of this in the field of education, for example, may be discovered in the Government School Inspector's Reports, while the frustration felt by the coal companies led them to demand the reorganization of local government itself.

Many kinds of problem were posed by the development of the Dukeries coalfield. How did religious institutions react to the arrival of the new population? Leisure presents a subject for study, not only because of the general interest of the manner in which miners and their families passed their 'spare' time outside working hours, but because here too formal facilities were slow to appear in the colliery villages, and when they did, they were very much in the gift of the coal companies and under the leadership of their personnel. Any study concerned with the history of coal-mining communities in the twentieth century should also approach the questions of the impact of the Second World War and of the nationalization of the mines on the villages. Did the coal companies lose control of their men during the war, and what kind of authority replaced theirs on their departure in 1947?

The development of new coalfields is inherent in the nature of coal-mining as an extractive industry, reliant upon the exploitation of new seams as old mineral resources are exhausted. Parallels with other developing fields may therefore be sought, and attention should be paid not only to what is known of the early history of other coalfields in Britain in the nineteenth and twentieth centuries, but to more recent cases such as that of the proposed Vale of Belvoir field. If lessons can be drawn from the common problems of new coalfields, the course of an industry which will offer an essential source of energy for the foreseeable future may more easily be planned. Nor is there any need to confine the analysis of new mining communities to this country. The company towns of the United States and the compounds of the mines of Southern Africa both offer interesting comparisons with the Dukeries villages. Moreover, it should be considered to what extent all new communities face similar problems and demonstrate common social phenomena, whether they be colliery villages, nineteenth-century factory villages or one-industry 'boom towns', or twentieth-century new towns, or housing estates planned 'from scratch'. The question does arise of the value of all 'model' communities built by a single authority, as opposed to the alternative policy of the renovation and development of existing towns and cities.

Little space in this book is devoted to that central part of the miners' lives which concerned work underground. This is not because the conditioning influence of the labour process is to be ignored. Indeed it is to be stressed that the Dukeries mining villages were functional, occupational communities which owed their very existence to developments in the industry. The shared experience and all-pervading influence of the job of producing coal had an immeasurable impact on the lives of the inhabitants. Nor is it sufficient to claim that this book is simply a social, not an economic, history. A denial of the influence of work and of labour relationships on the social structure of the villages would be wholly misguided. But it should perhaps be pointed out that the *peculiarities* of the new Dukeries coalfield, as compared with other coalfields in Nottinghamshire and beyond, seem to derive not so much from differences in the work process itself as from the monopoly of employment commanded by the coal companies in the model villages, and the power of the threat of dismissal; and also the unusually complete extension of the authority of the employer into village society and politics.

If the history of the Dukeries coalfield between 1913 and 1951 may be seen as a case study in the social and political problems of new coalfields in particular and all new communities in general, one may be grateful for the quality and range of evidence that enables most of the questions which suggest themselves to be answered. For a start, the period is recent enough to be within the memories of many still alive, not only miners but their wives and families, and residents of the 'old villages' as well. Their oral evidence should not be wasted, and while of course it should be granted the same exhaustive criticism as other forms of information, even subjective views and interpretations may help to create an understanding of the ethos of Dukeries society in the inter-war years. A formidable range of written evidence is also available: local newspapers; trade union records—both of the Spencer union and the 'official' NMA; company and business archives; local and central government material such as local authority committee minutes, electoral registers, and schools inspectors' reports; and the evidence still to be found at the pits themselves in the form of miners' record cards, signing-in books and leavers' registers. The industrialization of the Dukeries and its development as one of the most profitable and productive coalfields in Britain is a subject worthy of the most serious consideration, and it is fortunate that the tools are available to enable that task to be undertaken.

1. Development

'They struck coal in the middle of Saturday night . . .'[1]

The history of the Dukeries coalfield will be approached primarily as an example of the social development of new communities rather than as an economic or technical study of the exploitation of coal resources. Nevertheless, a brief description of the main stages of the sinking process is necessary in order to set the colliery villages in their industrial context. After all, the very reason why the Dukeries coalfield was not exploited until the 1920s is based in geology. The great coalfield of Yorkshire, Derbyshire, and Nottinghamshire, which has come to dominate the British industry's manpower and coal production figures in the twentieth century,[2] is composed of coal measures which outcrop on the surface at their western extremes, roughly a twenty-two-mile wide belt from Leeds in the north to Nottingham in the south. This coal was naturally the most easily won and therefore the first to be exploited, and coal-mining has been known in the 'exposed' coalfield since medieval times. However, the coal seams are tilted and dip gradually further and further beneath Permo-Triassic and Jurassic age rocks as one proceeds eastwards (see Fig. 2).[3] After the surface outcrops and the shallow coal had been extracted, the deeper seams were tapped, each new operation requiring a greater capital investment and a higher level of technological expertise as mining moved in an eastward direction. In the late nineteenth century the coal companies reached Nottinghamshire's Leen Valley; around 1900 they penetrated further east still, to develop the districts round Mansfield in Nottinghamshire and Doncaster in Yorkshire.[4]

[1] Interview, C. H. Green, Ollerton. See below, p. 17.

[2] See above, Introduction, Tables 1–2.

[3] The best technical account of the 'exposed', 'old concealed', and 'new concealed' coalfields is to be found in the Ministry of Fuel and Power's *North Midland Regional Survey Report* (HMSO, 1945).

[4] A. R. Griffin, *Mining in the East Midlands* (London, 1971), p. 160.

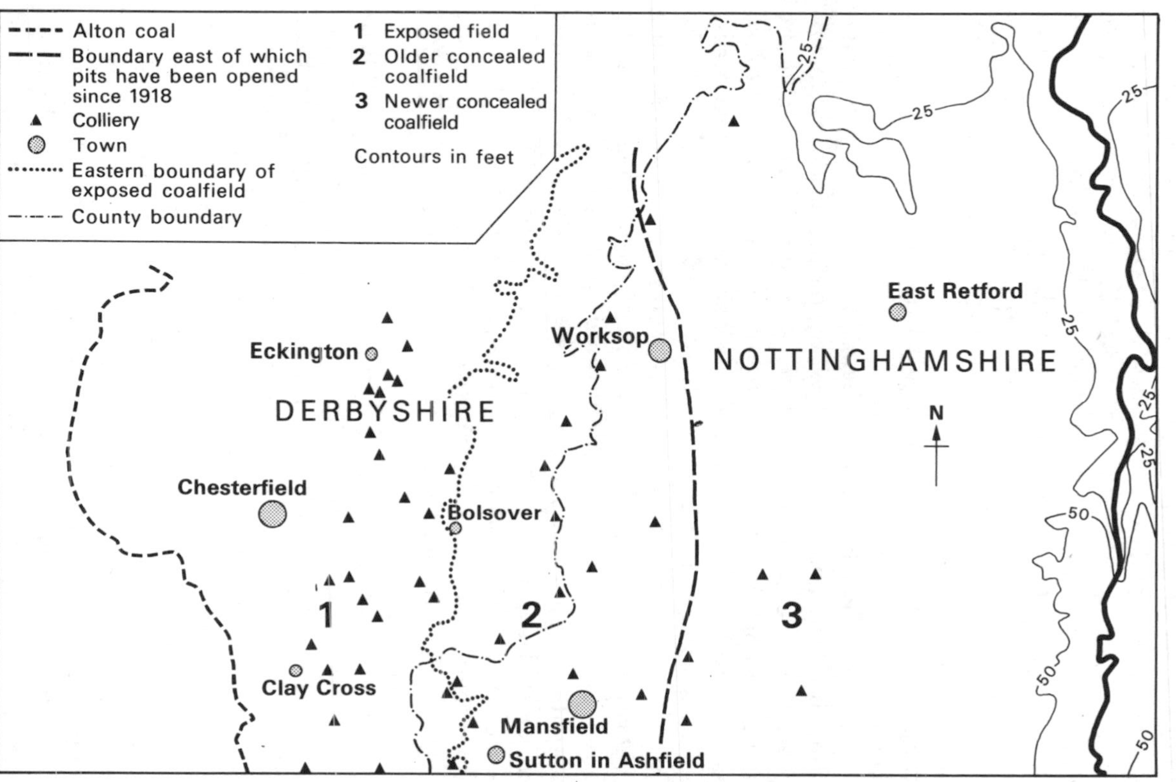

Fig. 2 Physical Map of the Nottinghamshire and Derbyshire Coalfield

Borings in the early years of the twentieth century proved that further rich seams existed at an even deeper level in eastern Nottinghamshire beneath Sherwood Forest and the area known as the 'Dukeries', shelving to an unreachable 4,000 feet beneath Lincoln Cathedral. The Dukeries coal was clearly a resource which would be well worth exploiting, but one which would require a heavy investment and advanced sinking techniques. The Top Hard seam, for example, which was particularly suitable for industrial rather than household purposes, would be struck at least 1,500 feet beneath the Dukeries.

The new field was situated beneath a part of Nottinghamshire hitherto almost untouched by industrial development, its surface being divided between agricultural land and the remains of Sherwood Forest. Landownership was concentrated in the hands of the aristocrats who had given the Dukeries its popular name: these included the Welbeck estate of the Duke of Portland, the Clumber estate (Duke of Newcastle), the Thoresby estate (Earl Manvers), and the Rufford estate (Lord Savile). The district was thinly populated, with a few villages actually situated on the estates, such as Thoresby's Budby model village, and some larger villages, often old market or coaching centres, which were now almost entirely dependent upon agriculture and the Sherwood Forest tourist trade. It was in this environment that the new coal mines of the 1920s were to be sunk.

Three of the new pits were in the Maun Valley, north-east of Mansfield. These were Clipstone, four miles from Mansfield; Thoresby Colliery at Edwinstowe, eight miles from Mansfield; and Ollerton, just over ten miles north-east of Mansfield. A colliery was sunk at Blidworth, five miles south-east of Mansfield, and one at Bilsthorpe, nine miles due east of the town. Further north, two collieries were sunk which looked not to Mansfield, but to Worksop or even into Yorkshire. These were Firbeck Main with its colliery village of Langold, six miles north of Worksop, and Harworth Colliery, with its colliery village of Bircotes, nine miles from Worksop, Retford, and Doncaster alike. Although both Mansfield and Worksop had been centres of mining for many years, the new pits were situated in open countryside previously untouched by coal-mining, isolated by poor bus services and a lack of private transport, and with traditions and customs far removed from those usually associated with established coalfields. Advancing technology had brought the prospect of a social and economic transformation in eastern Nottinghamshire.

There is no doubt that the exploitation of a concealed coalfield involved a considerable financial gamble. In the case of Kent, developed more or less at the same time as the Dukeries, the field never became profitable and to a large extent failed to live up to the high hopes it originally aroused.[5] Of the Nottinghamshire pits that were actually sunk in the 1920s (and some projects never progressed beyond the earliest stages of planning), there was one failure and one near failure. At Firbeck Main Colliery near Worksop it was expected that coal would be reached at a depth of 480 yards, but in fact workable seams were not struck until 880 yards, some 380 yards deeper than at the much older pit of Shireoaks which was only four miles distant. Then it was discovered that the shafts at Firbeck had been sunk in a fault area, and some very steep inclines had to be negotiated. Many other difficulties were encountered, and in 1938 the south working met a washout (a place where a prehistoric river had washed away the coal-producing vegetation). If it had not been for the increased demand for coal during the Second World War the colliery would have been closed, but it struggled on until 1968—thus forming another example of a phenomenon known in Kent, a colliery opened and closed within the memory of many of its workmen.[6] At Blidworth too, a fault was struck, and the colliery was closed between 1929 and 1932, leaving the village a ghost town, and threatening the investment of the Newstead Company.

There was, then, certainly a risk involved in sinking capital into a new and unproved coalfield. However, where the gamble was successful, and the rich seams of Top Hard materialized as expected, great financial rewards could be reaped by the entrepreneur. The coal companies did not draw

[5] See below, ch. 11.

[6] Correspondence, E. H. Baker, formerly Chief Clerk of Firbeck Main Colliery.

back from the attempt to exploit the new field. Indeed there was something of a 'scramble for the Dukeries'.[7]

For example, when Earl Manvers announced his intention of auctioning the mineral rights to the coal situated beneath his Thoresby estate in May 1919,[8] many large coal companies were interested. Often these were the mining branches of major Derbyshire-based iron- and steelworks. The Staveley Company, the Bolsover Company, the Stanton Company, the Sheffield Company, and the New Hucknall Company all made offers in addition to that of a consortium headed by a Mr Laverick. Sometimes coal companies joined forces to make preliminary investigations into the workability of exploiting the Dukeries field. Butterley and Stanton had become partners in a coal-proving venture at Ollerton in 1917, and early in 1920 the managing director of the New Hucknall Company suggested that all three concerns should merge operations to form a single unit.[9] However, this suggestion was rejected, and in May 1921 the Stanton Board notified Butterley that they wished to withdraw from the joint venture, whereupon Butterley took over Stanton's half of the lease from Lord Savile and repaid Stanton's outlay. According to the minute books of the Managing Committee of the Stanton Company, their option in the development of Ollerton Colliery was relinquished because the minimum mineral rents could not be less than £4,000 per annum if coal was proved at Ollerton, and they were likely to have to pay £5,000 a year for their solo development at Bilsthorpe.[10] They chose to give up Ollerton, which a majority of the committee thought would prove less profitable, although mining agent J. A. Houfton thought otherwise.[11] Moreover, it was probably thought that the *operation* of a pit as a joint venture was less satisfactory than the *proving* of coal, with its possibility of a financial loss which might be shared between companies.

[7] 'A Correspondent', 'The New Coalfield in Nottinghamshire', *New Statesman*, 24 Dec. 1927.

[8] Manvers Papers Ma 2C 132, Hewitt to Johnson, Raymond-Barker, 27 May 1919. For Manvers's decision to lease the mineral rights see below, ch. 3.

[9] A. E. G. Harmon, *Stanton Company History* (privately circulated, 1960, 7 vols.), vol. 3, p. 114.

[10] Stanton Company Managing Committee Minutes, 8 Aug. 1921.

[11] Stanton Company Board Minutes, 26 June 1920.

In the end, all but one of the new pits was singly owned: Ollerton by Butterley, Bilsthorpe by Stanton, Thoresby and Clipstone both by Bolsover, Harworth by Barber-Walker, and Blidworth by Newstead, a subsidiary of the Staveley Company. Firbeck Main was sunk by a union of the Sheepbridge iron and steel company and the Doncaster Collieries Association (another Staveley subsidiary). Firbeck, like Harworth, is something of an exception in that it is not in the Dukeries but geologically part of the South Yorkshire field. Although both pits are situated just within Nottinghamshire, Firbeck was in the South Yorkshire Area of the NCB and the Yorkshire NUM, while Harworth miners considered Doncaster as 'town', where commercial and recreational facilities might be sought.

In addition to this, work at Harworth started rather earlier than was the case at the other pits on the new coalfield, before the First World War. The financier and Conservative MP, Arnold Lupton, set about securing leases from existing landowners, and by 1910 had options on twenty-two areas covering several thousand acres. For two years he sought a colliery owner willing to finance his plan, but he was unsuccessful with all British companies, and in 1913 he approached the German mining magnate Hugo Stinnes, who put up sufficient capital to promote the Northern Union Mining Company, which came into being in August 1913. Immediately a German specialist sinking company sent over officials, workmen, and equipment to clear a site, and they were thus occupied when war broke out in August 1914. The German employees were interned after a combined raid by policemen and Sherwood Rangers, led by the Chief Constable, Captain Jamesson, and Colonel Whittaker, arrested sixteen German officials.[12] The properties of the company were impounded by the government under the Trading with the Enemy Act of 1916 at the instigation of the local aristocrat, Lord Galway of Serlby. Despite Lupton's protests, which included chaining himself to the railings in Downing Street,[13] work ceased until 1917 when a British Harworth Main Colliery Company was formed, initially with a capital of 100 one-pound shares, but

[12] Cyril Murphy, *Harworth Colliery 1924–74*, typescript, no pagination, Harworth County Library.

[13] Galway of Serlby Papers 13009–14.

increased in December 1918 to £80,100 in one-pound shares. The Barber-Walker Company had a controlling interest and took over the sinking of the mine themselves in April 1922.[14] At all the other collieries in the new field, coal was proved during the First World War and sinking operations were commenced shortly afterwards, mostly in the 1920s.[15]

Often the companies brought in personnel experienced in colliery development to supervise the sinking. L. T. Linley, who became the first manager of Bilsthorpe Colliery in October 1924, had previously been the manager of another Stanton Company pit at Pleasley in Derbyshire.[16] Several officials who were involved in the Bolsover Company's sinking operations at Clipstone moved directly on to Thoresby two or three years later.[17] At Langold, where the Sheepbridge Company and the Doncaster Collieries Association had combined for the purposes of sinking the new pit in 1915: 'The Sheepbridge Company provided the Administrative and Engineering Officials while the Doncaster Company provided the Mining Officials, each group of officials bringing friends, relatives and associates to form the main part of the workforce.'[18]

The development of all the new pits followed a similar pattern. The site had to be cleared, which often, as at Thoresby, involved the felling of old oak trees to make a clearing in Sherwood Forest.[19] Mineral rail lines had to be put in, temporary offices built, and then the shafts sunk, usually by the Belgian François cementation process.[20] After an anxious period of uncertainty, coal would be struck as the shafts reached the coal measures. The flags would then be hoisted. As one contemporary observer noted:

Three months ago the main seam of the Barnsley bed of coal—the finest coal in the world—was struck at Bilsthorpe pit in the new Nottinghamshire coalfield. Soon Bilsthorpe will be flinging another million tons of

[14] G. C. H. Whitelock, *250 Years in Coal, the History of the Barber Walker Company* (Derby, 1955), p. 166.

[15] See above, Table 3.

[16] Stanton Company Colliery Committee Minute Book, Minute 29, Oct. 1924.

[17] Thoresby Colliery Manager's Annual Report, 1925.

[18] Correspondence, E. H. Baker, Langold.

[19] Thoresby Colliery Manager's Annual Report, 1925.

[20] Ibid., 1925–8; Butterley Company H3/6 225.

coal a year into a market which, on the gloomy testimony of so many experts, is already so disastrously overstocked. To celebrate the event the Union Jack was swung high over the headgear at the pit! And possibly the demonstration was not altogether disproportionate to the hopes for the future of the British coal industry, which this new coalfield has legitimately aroused.[21]

The moment when coal was struck at another colliery, Ollerton, is still remembered:

In fact I came over one weekend, and I can still remember it quite well, and my mother said, your father's on twelve hours, are you going to take his dinner across? I should be about fifteen then, so of course I wouldn't be out for a drink, and when I got across my dad says to me, hey up, he says, they struck coal last night, he says, should you like to go down this afternoon with me and have a look at it? I says no, I'm not bothered, I don't want to go down the pits. He says, well, I've never been down one, so, he says, go down with me, the deputy's taking us down, a chap called Arthur Eyre. And I went down on the Sunday afternoon, that they struck coal in the middle of Saturday night and it were August 31st, 1925, and I don't think there's many living now that actually went down right from them striking coal, actually because most of the men then were middle-aged men, and if there's any living today, they'd be very, very old.[22]

Sinking was a lengthy process—it usually took around three years to reach the deep coal measures. At Ollerton, sinking had commenced in 1922. At Bilsthorpe, sinking started in autumn 1924 and coal was reached on 17 August 1927. At Thoresby, the last of the pits to be developed in the Dukeries in the years before the Second World War, sinking began in spring 1925 and coal was reached on 21 January 1928.[23] Even then the pits would not achieve their full production level for some years. Thoresby, for example, produced only 2,147 tons of coal in 1928, 43,963 tons in 1929, 278,543 tons in 1930, and achieved 668,662 tons in 1931. Thereafter it gradually increased to its pre-war record of 766,800 tons in 1937.[24] Similarly, Ollerton Colliery built up production slowly to reach its highest pre-war figure of 655,360 tons in 1937.[25]

[21] *New Statesman*, 24 Dec. 1927.
[22] Interview, C. H. Green, Ollerton.
[23] Thoresby Colliery Manager's Annual Report, 1928.
[24] Ibid., 1928–37.
[25] Butterley C4/2 33.

Table 5. *Bilsthorpe Colliery, month by month expenditure*[a]
(£)

	total expenditure	housing	shaft sinking
1925			
Jan.	9,069		
Mar.	29,491	13,574	
Apr.	42,985	16,574	
May	53,464		
June	70,710		
July	86,468	22,866	7,364
Aug.	109,540	25,579	12,585
Sept.	130,221	28,640	15,410
Oct.	154,666		
Nov.	175,696	32,660	21,898
Dec.	216,016		
1926			
Jan.	233,052		
Feb.	253,770		
Mar.	284,043	53,975	47,845
Apr.	300,996		
May	321,171		
July	356,633		
Aug.	387,676		
Sept.	387,676		
Oct.	411,647		
Nov.	435,908		
Dec.	466,079		
1927			
Jan.	483,351	111,871	159,534
Mar.	528,667		
Apr.	546,924	137,798	
May	576,129		
June	595,487		
Aug.	657,950		
Sept.	683,352	183,983	199,308
Oct.	703,213		
Nov.	714,878		
Dec.	730,247	192,767	205,368
1928			
Jan.	744,740		
Feb.	759,780	201,437	
Mar.	781,388		
June	805,996		

	total expenditure	housing	shaft sinking
July	821,068	205,072	
Aug.	841,652		
Sept.	851,090		
Oct.	866,030		
Nov.	874,082		
Dec.	887,187	225,040	
1929			
Nov.	916,555		
1930			
Apr.	956,014	233,902	225,183

[a] Source: Stanton Company Colliery Committee Minutes, 1925–30.

All this meant that a vast expenditure was incurred well before any profit might be expected to be made. Consider the month-by-month expenditure of the Stanton Company during the early years of the development of their Bilsthorpe pit, as shown in Table 5. It will be noticed that a major element of the expenditure incurred involved the erection of the colliery village, with its workmen's, officials', and manager's houses, shops, institutes, and village hall.

A complete breakdown of the capital expenditure at Bilsthorpe to April 1930 (see Table 6) indicates the complexity of the tasks involved in the sinking of a new colliery. However, even this does not represent the full extent of the cost of sinking a new pit to the entrepreneur. There should be added the coal royalties paid to Lord Savile, on whose estate the pit was sunk. Seventy thousand pounds of minimum rents had been paid by May 1926, still more than a year before coal was reached, and after the mine became productive the royalties amounted to approximately £10,000 a year. This was typical of all the Dukeries pits,[26] and did not include the cost of any surface land leased or bought for the purpose of erecting a colliery village; at Bilsthorpe the Stanton Company bought 100 acres from Lord Savile at £154 an acre in 1925.[27]

[26] See below, pp. 67–71

[27] Stanton Company Colliery Committee Minute 86, 9 Feb. 1925.

Table 6. *Bilsthorpe Colliery, expenditure to 30 April 1930*[a]

	(£)
Clearing site	1,169
Cart roads and pit yards	2,838
Railway	57,853
Locos	5,284
Ambulance room and garage	1,304
Loco sheds, shops, stores	18,001
Colliery water supply	11,674
Reservoir and water cooling	3,949
Colliery drainage	5,211
Boiler plant and steam mains	31,708
Power house and plant	26,042
Colliery electrical mains	4,051
Winding engines and capstans	28,341
Fan house and plant	5,833
Head gears	7,221
Sinking to Top Hard seam (both shafts)	207,491
Sinking to Low Main seam (no. 1 shaft)	17,698
Shaft equipment	28,008
Banking plant	4,342
Screening plant	29,413
Coal washing plant	34,872
Loose tools	1,711
Sundries, administrative expenses	19,368
Temporary plant	23,740
Brickyard capital	7,835
Workmen's Institute	1,147
Repairs	12,843
Garage for motor lorries	483
Aerial ropeway	5,428
Compressed air plant	2,386
Offices	2,313
Pit bottom equipment (conveyors)	105,379
Lamp room	4,481
Pit tubs	6,213
Housing	223,902
Total	956,014
Estimated further expenditure	38,040
	994,054

[a] Stanton Company Colliery Committee Minutes, 16 June 1930.

There were many other essential arrangements which had to be made before the pit could reach fully productive status. Railway lines had to be constructed almost before any work could proceed, in order to ferry sinking-equipment to the site as well as to take any coal produced. However, the railway companies were as cautious as the coal companies in committing themselves to expenditure before the new coalfield had been proved a success. The Stanton development at Bilsthorpe may be taken again as one example. As early as January 1919 the Stanton Company had decided that sinking at Bilsthorpe should be proceeded with at once, providing that a railway company would furnish the necessary temporary facilities for bringing materials to the site.[28] But in March of the same year, Stanton were to be found opposing the Mansfield Railway Company's proposed bill in Parliament, as they had not agreed on the standard rate for haulage for the three mile link to Bilsthorpe from the existing terminus of the mineral line at Rufford (a pit sunk during the First World War). Stanton held out for a rate of 1*d.* a ton.[29] As a result the Mansfield Company did not construct a railway line to Bilsthorpe, and it was not until November 1922 that the Midland Railway Company agreed to build a contractor's line to Bilsthorpe at Stanton's expense. Should the pit prove profitable and a permanent line be required, Midland would pay for this themselves.[30] However, negotiations over the contractor's line were still in progress in March 1923,[31] and at the Managing Committee meeting of 9 April 1923, it was decided that as no progress was being made with the Midland line the matter should be taken up with the Great Central Railway. This threat did stimulate the Midland (LMS) to agree to start work on the contractor's line.[32] But a further delay was encountered when one of the landowners whose property was required for the railway line, W. Wilson, refused to sell his land to Stanton for £150 an acre. He tried instead to lease the land, claiming a minimum rental of £400 per annum for the coal royalties.[33] This meant that the line had

[28] Stanton Company Managing Committee Minutes, 14 Jan. 1919.
[29] Ibid., 11 Mar. 1919.
[30] Ibid., 27 Nov. 1922.
[31] Ibid., 12 Mar. 1923.
[32] Ibid., 14 May 1923.
[33] Ibid., 8 Oct. 1923.

to be diverted from its original route passing over Wilson's land to one passing over Lord Savile's, for the latter was prepared to sell at a rate of £100 per acre (arable) and £75 per acre of woodland.[34]

Nevertheless, this substantial investment necessary for the exploitation of a new coalfield was regarded by the coal companies as well worth while. In the inter-war years the rich, expanding Dukeries coalfield offered a great opportunity in a declining industry negotiating the perils of falling domestic consumption and export orders and the political hazards of attempting to reduce wage costs. By the time the pits reached full productive capacity in the 1930s, they were making substantial profits for their owners, and beginning to repay the sums invested in their sinking. In 1937, for example, Clipstone Colliery made a profit of £195,356 for the Bolsover Company, while Thoresby was in the black to the tune of £188,620. These were by far the best returns of any Bolsover pit.[35] The companies operating in the new field were amongst the most successful and productive concerns in the British coal industry. The Samuel Commission found that in 1925 the Butterley Company's trading profit was £239,390, and that for the Staveley Company in 1924 was £365,537.[36]

When the Reorganization Committee of the early 1930s investigated possibilities of amalgamation (a plan vigorously opposed by the Nottinghamshire companies), they found that 49 per cent of the coal production of the Notts.-Derbys. field in 1932 was achieved by the seven largest companies. These included five of the concerns which had sunk new pits in Nottinghamshire in the 1920s: Bolsover (3,295,000 tons in 1932), Butterley (2,176,000), Staveley (1,664,000), Stanton (1,609,000), and Barber-Walker (1,084,000).[37]

It is scarcely surprising that the coal companies which developed the new Nottinghamshire coalfield were in general large and efficient firms. They had to be capable of absorbing the considerable cost of sinking a deep mine in the new concealed field, and they had to be able to outbid possible rivals in order to secure the mineral rights from the local landowners.

[34] Ibid., 14 Jan. 1924.

[35] Bolsover Company Annual Report, 1938.

[36] Report of the Royal Commission on the Coal Industry 1925, PP 1926 xiv, 109.

[37] PRO COAL 12/144.

In the 1930s the companies operating the Dukeries pits were amongst the greatest coal producers in the country. By the Second World War, the Bolsover Company was the third largest colliery concern in Britain.[38] The Staveley-Sheepbridge group, responsible for sinking Blidworth and Firbeck Collieries, was producing 19,650,000 tons a year in 1942, a greater output than any other British colliery undertaking.[39] The Stanton Ironworks Company secured one of the great successes of the inter-war British iron and steel industry when they secured the British licence for the Delavaud process, which gave them a financial security which was probably responsible for enabling them to work Bilsthorpe Colliery full-time throughout the inter-war period.[40]

Nor was there any end in sight in the foreseeable future to the profitable reserves of the Dukeries field. In 1945 the life of the Low Main coal seams was officially estimated as 'at least 141 years'.[41] A less conservative estimate in 1961 suggested that at present production levels Bilsthorpe Colliery could continue production for at least 480 years and Ollerton Colliery for at least 313 years.[42]

As it happened, although the expectation still remains that the Dukeries coalfield will continue to produce profitable coal for another century at the minimum, the coal companies which took the risk of investing in new pits in Nottinghamshire in the 1920s never did fully recoup their outlay. This was not for technical reasons but because of political change —the nationalization of the British coal industry after the Second World War. This should help to explain why the coal companies were so interested in what went on in the colliery villages outside working hours. They interpreted the need to secure continuity of production and the need to protect their investment in a wide sense, and rightly so: the main threats to the success of their venture in opening up the Dukeries coalfield lay in politics and industrial relations. As far as a strictly unrestricted market was concerned, the Dukeries pits

[38] J. E. Williams, *The Derbyshire Miners* (London, 1962), p. 163.

[39] A. R. Griffin, *Mining in the East Midlands*, p. 163.

[40] Information, Dr S. D. Chapman, Nottingham University.

[41] *North Midlands Regional Survey Report* (HMSO, 1945), p. 18.

[42] C. M. Law, 'Aspects of the Economic and Social Geography of the Mansfield Area', Nottingham University M.Sc. 1961, p. 78.

came cheap, even at a million pounds each. They could be described as logical technical developments to take the coal industry into the twentieth century. Unfortunately for the Dukeries coal companies, that century also brought the end of private enterprise in the coal mines of Britain.

2. Migration

'There is a Dukeries coalfield; there is as yet no Dukeries miner'[1]

I

It is well known that a large-scale migration of labour has taken place as a result of the overall decline of the British coal industry since the end of the First World War. In 1923 over a million men were employed in the pits;[2] by 1980/1 the figure was a little over 220,000.[3] Certain social effects of this dramatic industrial change have been studied. The influx of workers from the South Wales coalfield to 'new industry' towns like Oxford, for example, provoked contemporary analysis and more recent work.[4] Correspondingly, the effects of pit closure on the coalfield communities themselves have attracted attention. J. W. House and E. M. Knight considered the case of Durham in the 1960s, as did the Department of Employment and Productivity in *Ryhope, a Pit Closes.*[5] Martin Bulmer discussed the declining mining town of Spennymoor in a collection of essays entitled *Mining and Social Change, Durham County in the Twentieth Century* (London, 1978). Knight on his own dealt with the case of West Cumberland.[6] In general, the conclusions of these studies show the expected dislocation in the society as well as the economy of the declining coalfields, although Bulmer in particular

[1] *New Statesman*, 24 Dec. 1927.

[2] 1923 UK total: 1,203,384 (PP 1937-8 xxvii, pp. 332-3).

[3] 1980/1 UK total: 224,841 (NCB Annual Report, 1980/1, p. 22).

[4] H. Makower, J. Marschak, and H. W. Robinson, 'Studies in the Mobility of Labour—a Tentative Statistical Measure', *Oxford Economic Papers*, 1 (1938), 83-123. R. C. Whiting, 'The Working Class in the New Industry Towns between the Wars, the case of Oxford', Oxford University D.Phil. 1978. See below, p. 287.

[5] J. W. House and E. M. Knight, 'Pit Closure and the Community', University of Newcastle upon Tyne *Papers on Migration and Mobility in Northern England*, 5 Dec. 1967. *Ryhope, a Pit Closes* (HMSO, 1970).

[6] E. M. Knight, 'Men Leaving Mining', University of Newcastle upon Tyne *Papers in Migration and Mobility in Northern England*, 7 Jan. 1968.

stresses the durability of the tightly-knit social structure of the colliery village, even after the closure of the pit and the transfer of much of its labour to factory development or even to the ranks of the unemployed.

This concentration on the overall picture of mining as a declining staple industry has, however, masked another important aspect of migration. Not only were men leaving the coalfields in the 1920s and 1930s, but they were moving within and between coalfields, particularly when new pits were opened. The clearest (though not the most extensive) example of a new coalfield being exploited at this time, that of the Kent field, has been considered by W. Johnson[7] and R. Goffee.[8] But only four pits were in operation even at the peak of Kent's development, while its physical isolation from traditional mining areas makes it something of an unusual case. Let us consider, rather, the Dukeries coalfield of Nottinghamshire, opened up in the 1920s. The theory that just as much social dislocation was consequent upon the opening of a pit as on a closure may be examined.

While it is clear that oral evidence is essential in helping to explain the motives of the miners in moving to the new field and their reactions to it, electoral registers, colliery and company archives, and the records of schools should be used to give a statistical basis for the study of the numbers of people migrating, where they came from, how long they stayed, and where they went, and something of their age and marital status. The degree of migration and mobility is an important indicator of the fluidity or stability of any community. If the level of migration is found to be high, it would suggest that the traditional picture of mining communities typified by a close-knit society enjoying little geographical mobility might require substantial revision in the case of new coalfields.

If the history of the Dukeries coalfield is to be viewed as a case study on the nature and effects of economic, social, and political change, it will be useful to start with an analysis of

[7] W. Johnson, 'The Development of the Kent Coalfield, 1896–1946', Kent Ph.D. 1972.

[8] R. A. Goffee, 'Kent Miners: Stability and Change in Work and Community', Kent Ph.D. 1978.

the migration involved in that change. As J. A. Jackson put it:

The movement of population has been and remains an essential component of economic development, social change and political organisation. Capital investment must be matched by investment in labour resources . . . the growth of cities, the development of new resources and territories and the increasing international content of business, leisure and political experience depends upon the settlement, temporary or permanent, of individuals in diverse locations away from their place of birth and upbringing.[9]

II

First of all the numbers involved in the migration to the new Nottinghamshire field must be estimated. As Table 7 shows, the population of the new colliery villages constructed in the 1920s increased by some 25,000 net. These increases are also reflected in the rapidly growing population of the local authorities in which the new pits were situated, Southwell Rural District and Worksop Rural District, whose gain by migration is contrasted in Table 8 with the loss experienced by older coalfields at this same time. These increases by migration were due to the transformation, shown in Table 9, of the economic base of Southwell and Worksop Rural Districts.

Table 7. *Population of new colliery villages*[a]

village	1921	1931	colliery
Bilsthorpe	134	1,972	Bilsthorpe
Bircotes	865	6,092	Harworth
Blidworth	2,033	5,316	Blidworth
Boughton	315	1,318	Ollerton
Clipstone	592	3,443	Clipstone
Edwinstowe	963	2,818	Thoresby
Langold	232	4,307	Firbeck
Ollerton	676	3,912	Ollerton
Rainworth	342	894	Rufford

[a] Source: 1931 Census, Nottinghamshire, Table 2, p. 1.

[9] J. A. Jackson, 'Migration–Editorial Introduction' in J. A. Jackson (ed.), *Migration* (Cambridge, 1969), p. 1.

Table 8. *Population of local authorities 1921-31*[a]

	1921	1931	% change	births	migration
Southwell RDC	20,157	32,076	+59.1	+9.8	+49.3
Worksop RDC	5,070	14,555	+187.1	+27.8	+159.2
Mansfield Woodhouse UDC	13,477	13,721	+1.8	+12.6	−10.8
Huthwaite UDC	5,478	5,092	−7.0	+7.8	−14.8
Bishop Auckland UDC	14,290	12,277	−14.1	+6.6	−20.7
Spennymoor UDC	18,238	16,369	−10.3	+10.9	−21.2

[a] United Kingdom Census, Notts. and Durham, 1931.

Table 9. *Numbers occupied in mining 1921-31*[a]

	no. of miners		% of all workers	
	1921	1931	1921	1931
Southwell RDC	327	4,092	4.9	39.37
Worksop RDC	378	3,408	14.9	71.46

[a] UK Census, Notts. pp. 4-7, Table 3, 1931.

Table 10. *Turnover of labour, Harworth Colliery*[a]

	no. of men signing on	total workforce	net gain
1925	992	1,143	
1926	812	1,364	+221
1927	784	1,123	−241
1928	unavailable	1,335	+212
1929	1,110	1,728	+393
1930	1,132	2,080	+352
1931	928	2,262	+182
1932	398	2,021	−241

[a] Mines Department, *Annual List of Mines*, 1925-32; Harworth Colliery signing-on books, 1925-7, 1929-32 (the volume for 1928 is missing).

However, this net gain of approximately 8,600 miners[10] and 25,000 population in the new coalfield between 1921 and 1931 does, in fact, conceal a much greater level of immigration, as Table 10 indicates. In each year up to 1931, well over half the workforce at Harworth was renewed each year—a formidably rapid turnover, and one which implies that many more men and their families must have passed through the new colliery villages than at first sight appeared. At Harworth alone, 6,156 *men* signed on in the seven years tabulated above, many of them married. Although it is true that Harworth was the largest and the most unstable of the new pits, its heavy turnover of labour was not unique in the coalfield, as the figures for Thoresby, Bilsthorpe, and Ollerton in Tables 11-13 indicate.

Table 11. *Workforce and turnover of labour, Thoresby Colliery*[a]

	men signing on	leaving	av. total workforce
1930	942	233	1,364
1931	670	607	1,464
1932	421	348	1,538
1933	213	274	1,478
1934	145	232	1,383
1935	365	273	1,450
1936	153	250	1,378

[a] Thoresby Colliery Manager's Annual Reports, 1930-6.

Another way of measuring the rapid turnover of population which affected the new communities is to examine electoral registers. In their early years, the electoral rolls showed a large number of new names in the colliery villages, as may be expected during a period of population expansion. But the turnover of voters continued even after the population reached a stable level. Table 14 compares a complete set of the voters in the various colliery villages, in a year soon after their numbers levelled off, with the electoral roll of a few years later, showing how many people stayed in

[10] Blidworth Colliery was at this time in Skegby RDC, so its workforce must be added to the figures in Table 9.

Table 12. *Turnover of labour, Bilsthorpe Colliery*[a]

	men signing on	leaving		men signing on	leaving
1929 Jan.	72	17	1930 Jan.	70	48
Feb.	87	21	Feb.	108	40
Mar.	87	21	Mar.	89	27
Apr.	108	45	Apr.	85	46
May	68	19	May	24	49
June	57	60	June	28	64
July	36	50	July	10	68
Aug.	31	17	Aug.	20	38
Sept.	55	54	Sept.	56	29
Oct.	99	36	Oct.	90	108
Nov.	87	33	Nov.	57	38
Total	818	400	Total	675	587
1931 Jan.	85	32	1932 Jan.	32	19
Feb.	95	19	Feb.	45	23
Mar.	103	52	Mar.	13	24
Apr.	102	38	Apr.	23	25
May	29	32	May	50	24
June	25	38	June	25	53
July	81	25	July	0	27
Aug.	89	13	Aug.	0	307
Sept.	70	21	Sept.	17	10
Oct.	89	35	Oct.	48	76
Nov.	54	24	Nov.	19	22
Dec.	8	13	Dec.	34	15
Total	830	342	Total	306	625

[a] Stanton Company, Bilsthorpe Colliery Manager's Monthly Reports, 1924–32.

the villages for a substantial period of time. The unusually low total of long-term residents at Blidworth is accounted for by the fact that in 1929 the pit was closed temporarily after a fault was struck, and most of the miners living there had to leave to seek other employment. Many did not return when the pit reopened. Annesley is included in Table 14 as an example of a stable nineteenth-century mining village in the older west Nottinghamshire coalfield, and Newark and Retford as non-mining districts in the same parliamentary

Table 13. *Men set on and workforce, Ollerton Colliery*[a]

	men set on	total employed
1925	258	327
1926	637	958
1927	1,053	1,399
1928	1,068	1,712
1929	993	1,838
1930	526	1,535
1931	585	1,733
1932	366	1,641
1933	406	1,550
1934	559	1,591
1935	292	1,604
1936	281	1,631
1937	294	1,643
1938	219	1,616

[a] Ollerton Colliery signing-on books, 1925-38.

Table 14. *Turnover of voters in the colliery villages*[a]

	total voters	% remaining
Ollerton	1929: 1,608	1938: 35
Harworth	1930: 2,385	1937: 29
Edwinstowe	1930: 1,068	1938: 34
Clipstone	1929: 1,875	1938: 35
Blidworth	1927: 1,681	1935: 9
Langold	1930: 2,139	1937: 36
Annesley	1922: 496	1930: 55
Newark (East Ward)	1930: 600	1935: 58
Retford (East Ward)	1932: 450	1936: 56

[a] Electoral registers, parliamentary constituencies of Newark, Bassetlaw, and Broxtowe, 1922-38.

constituencies as the new pits. Needless to say, the years chosen in Table 14 are somewhat arbitrary, and the number of voters remaining has probably been slightly underestimated due to the difficulty of identifying women who marry and change their name, although a comparison of male voters only in Edwinstowe suggests that this number is low. Table 15 shows internal migration; that is, the number of people changing address within villages, 'moving' rather than migrating. Those merely moving within the villages are included with those remaining as voters in Table 14 above.

Table 15. *Internal migration in the colliery villages*[a]

	%
Harworth, 1930–7	6.3
Clipstone, 1929–38	11.4
Ollerton, 1929–38	12.0
Bilsthorpe, 1931–8	10.8
Edwinstowe, 1930–8	13.0

[a] Newark constituency electoral rolls.

The evidence of the electoral rolls and the colliery signing-on books which suggests that many miners and their families were leaving the colliery villages during their early years is strengthened by other sources. For example, many of the admissions registers of the elementary schools in the new villages survive. These can be used to estimate turnover of population since they record the date of each child's departure and the reason for it. Table 16 shows the percentage who left before their transfer to an upper school (or before they reached the statutory leaving age), in order to move to another district. The last three schools are included in Table 16 as a comparison and a contrast—these were situated in towns on the older western coalfield. These were the communities from which many of the immigrants to the new pits came, so a striking conclusion may be drawn: proportionately more families *left* the expanding new villages than left declining mining communities. There is no evidence that miners with children were any more likely to leave the new pits than single men. Indeed the reverse was probably the case;[11] so, if anything, the schools admissions and leavers registers slightly

[11] See below, p. 35.

Table 16. *Percentage leaving school to move to another district*

		total leaving	all children	%
1925–32	Harworth[a]	882	1,810	49
1922–32	Blidworth[b]	194	459	42
1924–32	Langold[c]	576	1,891	31
1922–32	Ollerton[d]	176	615	29
1922–32	Kirkby-in-Ashfield[e]	210	1,099	19
1923–32	Sutton-in-Ashfield[f]	177	926	19
1924–32	Hucknall[g]	68	1,586	4

[a] Nottinghamshire County Records Office SA Schools Admissions Registers, Harworth Bircotes Junior Mixed Infants SA 15 1/1.
[b] Notts. CRO SA 17 2/1, 2/2 Blidworth Wesleyan.
[c] Notts. CRO SA 105/1/1, Langold Temporary.
[d] Notts. CRO SA 128, Ollerton Wellow Rd. Mixed.
[e] Notts. CRO SA 97 4/2, Kirkby-in-Ashfield East Council School.
[f] Notts. CRO SA 163 10/2, Sutton-in-Ashfield Central Infants School.
[g] Notts. CRO SA 91 1/4, Hucknall Upper Standard Mixed.

Table 17. *Length of stay or immigrant miners*

% left after	Bilsthorpe	Edwinstowe	Ollerton	Harworth
1 month(s)	2.1			
3 ..	5.8			
6 ..	12.0			
1 year(s)	20.4	28.1	31.3	27.0
2 ..	34.1	41.3	41.1	45.2
3 ..	45.5	46.1	46.6	51.8
4 ..	54.5	52.8	51.8	57.2
5 ..	60.5	59.8	54.0	62.4
6 ..	66.0	61.5	56.0	65.5
7 ..	69.6	65.2	58.9	68.8
8 ..	72.7	65.8	61.3	70.6
9 ..	75.8		64.8	
10 ..	77.6			

underestimate the labour turnover in the coalfield. The effects of this labour circulation on the education of the migrant children will be considered below.[12]

By using electoral registers and colliery signing-on and leavers' books it is possible (Table 17) to construct profiles of

[12] See below, ch. 7.

the typical length of stay of miners in a number of the new colliery villages. It is clear that there was a considerable level of migration taking place in the new Nottinghamshire coalfield in the 1920s and 1930s, both of men taking up employment *and* leaving—a turnover which indicates that far more people were affected by the problems of the new communities than the size of the villages indicates.

III

We must now turn our attention to considering where the immigrants came from. Once again we may use colliery signing-on books (which give details of where the men were last employed) and school admissions registers (last school attended). Table 18 takes the example of Harworth. The high figure for Yorkshire is accounted for by the fact that Harworth is only a couple of miles from the county boundary; the nearest pits to Harworth—Maltby, Rossington, Hatfield, Dinnington, and Thurcroft—account for 1,499 of the 4,884 miners who signed on at Harworth between 1925 and 1931. Similarly, the three Worksop pits, Steetly, Shireoaks, and Manton, provide 467 of the Nottinghamshire contingent. We should conclude therefore that most of the men put on by the Barber-Walker Company at Harworth Colliery last worked in the pits nearest to it. However, if the migration was primarily short-distance, we should not underestimate the break involved in moving to Harworth. In a time before many miners possessed motor cars, Harworth was remote enough to be beyond reasonable walking distance of even the nearest pit community. The raw company village of Bircotes, built by Barber-Walker for the Harworth miners, would present an unfamiliar aspect to the migrants, with its new housing, high population turnover, lack of facilities, and 'unneighbourliness'. Moreover, Harworth also attracted a considerable number of 'long-distance' migrants from far-flung coalfields throughout Britain, amounting to around 20 per cent of the total in Table 18. It was reputed to be the first pit on the road south for Durham miners seeking employment as collieries closed in their own coalfield. Lancashire and Staffordshire each provided substantial contingents as well.

Table 18. *Migration to Harworth 1925-31*[a]

origin	no.	%
Yorkshire	2,359	48.3
Nottinghamshire	949	19.4
Derbyshire	414	8.5
North East	412	8.4
Lancashire	350	7.2
Staffordshire	160	3.3
Wales	31	0.6
Kent	28	0.6
Leicestershire	24	0.5
North West	21	0.4
Scotland	16	0.3
Warwickshire	7	0.2
others	20	0.4
unidentified	93	1.9

[a] Harworth Colliery signing-on books, 1925-31.

What differences may be discerned between these men and those of more local origins? Table 19 shows that the long-distance migrants were older, more likely to be married, and more likely to live in Bircotes than to commute by bus or cycle from other villages. They were also slightly more likely to stay for at least two years—40 per cent of their children left school within that time compared with 49 per cent amongst the complete sample.[13] One reason for these differences is the great mobility shown by young, unmarried men who were quite prepared to move from pit to pit within the region, often leaving Harworth to work for a short period elsewhere before returning—about fifty of the 928 men signing on at the pit in 1931 had worked at Harworth before. It was mainly single men who lived in Rossington, Hatfield, Maltby, or Worksop and travelled each day to Harworth. There are several examples in the signing-on books of men reappearing several times, having 'done the rounds' of the pits. Individuals can also be found who worked at a number of the new Notts. mines; indeed, 20 per cent of the Notts. men who signed on at Harworth between 1925 and 1931

[13] See above, Table 16.

Table 19. *Long-distance and short-distance migrants to Harworth, 1927*[a]

	short-distance	long-distance
aged under 20	107	5
20-9	265	69
30-9	183	41
40-9	67	23
50-9	1	2
average age	27	31
number married	56%	66%
living in colliery village	60%	95%

[a] Harworth Colliery signing-on book.

previously worked at one of the other pits sunk in the 1920s.

Distinct patterns of migration emerge. When the Harworth pit was first sunk, many men came from Creswell in Derbyshire and Maltby, Yorkshire. Later in the development of Harworth, the most notable single pits were Rossington, Thorne, Hatfield, and Firbeck. There is little evidence that many migrants came from other Barber-Walker Company pits—only thirty from Bentley and nine from Watnall in the six years between 1925 and 1931—well under 1 per cent of the total. However, links were built up with donating pits for other reasons. The company organized a recruiting drive in the Wigan area in the 1920s—a district which was to account for nearly half the total Lancashire immigrants in these years.[14] Another link was forged with St. Hilda's Colliery, South Shields (fifty-eight men between 1925 and 1931), noted for its colliery band. There was a sudden influx of almost 200 men from Dinnington within two months in summer 1931, when that colliery laid off men. But more often, hearsay and family contacts seem to have been more significant in directing migration. This was certainly the case as far as Kent[15] and Oxford[16]

[14] Information, Alf Grainger, Admin. Officer, Harworth Colliery.

[15] R. Goffee, 'The Butty System in the Kent Coalfield', *Bulletin of the Society for the Study of Labour History*, 34 (1977), 41-55.

[16] H. Makower, J. Marschak, and H. W. Robinson, *Oxford Economic Papers*, 1 (1938), 83-123; Goronwy H. Daniel, 'Some Factors Affecting the Movement of Labour', *Oxford Economic Papers*, 3 (1940), 149-79.

were concerned in the 1920s and 1930s; and the predominantly local nature of pre-nationalization coal companies reduced their ability to organize recruiting in distant coalfields. Often men signed on in pairs, having previously worked in collieries as far afield as Northumberland. It is likely that they had taken the risk of coming south without any better security or knowledge of Harworth than rumour. This is borne out by the fact that as time went on the 'recruiting ground' for Harworth spread outwards, the initial labour coming from local pits, and the influx from Durham reaching its peak only after 1927.

The picture of migration to Harworth is reinforced if we consider the other new Nottinghamshire collieries of the 1930s. An interviewee who first came to Ollerton in 1927 demonstrates the individual decisions characteristic of the migration:

RJW Before you lived in Nottinghamshire, where did you originally come from?
JHG Pilsley, in Derbyshire, near Chesterfield.
RJW What made you decide to move from Derbyshire to Nottinghamshire?
JHG Well of course it were the old days when there weren't a lot of work knocking about and Pilsley pit was rocking, ready for closing. So I moved into the new coalfield.
RJW What made you decide to go to Ollerton rather than any of the other pits that were opening up?
JHG Well, I had a friend, a relative there. He persuaded me to go.[17]

This is an example of 'chain migration', the process by which one or two 'pioneers' arrive in a distant community more or less by accident, inform their friends and relatives back home, and encourage further migration from what could be as restricted a donating district as a single village.[18] For example a high proportion of Welsh migrants to Oxford in the 1930s came from the Maesteg area, and in particular Pontycymmer village. The 1938-9 survey of social services in the Oxford district, which described this phenomenon as 'lump' migration, claimed that it restricted assimilation into

[17] Interview, J. H. Guest.

[18] For further examples of chain migration see Charles Price, 'Assimilation' in J. A. Jackson (ed.), *Migration*, p. 210.

the new community, because the Welshmen preferred to mix with their own kind, and Welsh clubs were founded at Cowley and elsewhere.[19]

As at Harworth, about three-quarters of the immigrants at Ollerton were relatively short-distance, from the older coalfield in West Notts. and from over the county boundary in Derbyshire. Between 1922 and 1932 48 per cent of the immigrants to Ollerton came from Nottinghamshire, 25 per cent from Derbyshire, and 27 per cent from more distant coalfields.[20] Mining towns suffering pit closures, such as Heanor, Ripley, Codnor, and Alfreton in Derbyshire, provided many migrants, as did Kirkby, Sutton, and Mansfield in Nottinghamshire. Of the new arrivals 5 per cent had previously lived in Harworth, which could possibly have been a temporary halt on the journey south from Durham. Other new pits like those at Clipstone and Edwinstowe had donated more men, further evidence that it was quite common to move from pit to pit in the new coalfield.

Subjective evidence adds colour to the general impression gained of the origins of the immigrants from colliery and educational records:

RJW Where did the miners come from in the early days?

AEC Nearly always from the Derbyshire pits, particularly those that were worked out or coming to the end of their useful life . . . quite a lot as I recall from Ripley area and mines thereabouts: Marehay, which even then was closing; Denby Hall, and a lot of those collieries in that area.

RJW And there weren't many immigrants from further away?

AEC Yes, there were always some and they were outstanding. You would notice them immediately from speech and they didn't seem to be readily absorbed into the same little coteries or the same little groups in the pubs and so on. There were certainly Scotsmen here then, Welsh and others, but particularly what the mining company was looking for of course was men who were used to the same conditions, the heavy stone conditions that they would meet here.[21]

Rather more men came from other Butterley pits to Ollerton than migrated from Barber-Walker pits to Harworth:

[19] *Survey of Social Services in the Oxford District* (1938–9), vol. 1, p. 59.

[20] Ollerton Wellow Road and Whinney Lane Council Schools admissions registers; Ollerton Colliery signing-on books.

[21] Interview, A. E. Corke, Ollerton.

CHG Well, a lot came from Staffordshire, because I'll tell you, I was a member of the colliery band, and we got quite a few players from around Silverdale and Stoke, round that way. But actually a lot did, as well, they followed the bosses what come here, because it were a Butterley Company pit and a lot came from East Kirkby, which is only perhaps twenty miles away.

RJW And also were there a lot of people from Derbyshire?

CHG Oh, yes, yes, round about 1929, 1930 time there was a pit got flooded, Clowne Shireoaks . . . and of course they closed the pit down, and a lot came from there.[22]

The Butterley Company took an active interest in pit closures, and sought labour where the seams were known to be failing. In May 1939 they were looking for young men between the ages of 18 and 25 among those put out of work by the closure of Pendleton Colliery in Lancashire.[23] In June of the same year they were on the look-out for possible employees from the Cannock Chase area, and in 1940 a group of coal-face workers from Northumberland was set on.[24] Other colliery companies also looked for suitable labour in the older coalfields of Britain. In April 1929 the Stanton Company Colliery Committee, which dealt with Bilsthorpe pit, noted that:

As previously reported, we have had representatives in the various districts, i.e. Warwickshire, Leicestershire and Staffordshire, and they brought back a few more men. We are at a loss to know what else to do in this matter. I believe that when Harworth Colliery were short of men some time ago, they opened up Offices in various mining towns in Warwickshire, and other places, and signed men on there sending them straight to the Colliery.[25]

Langold, like Harworth, was situated almost on the Yorkshire border, and most children attending school in the colliery village came from the nearest pit communities across the county boundary near Doncaster at Dinnington, Thurcroft, Maltby, Grimethorpe, and Bentley. Most of those from Nottinghamshire were previously in the Worksop district, while those from Derbyshire were most frequently from Creswell.

At Blidworth similar proportions of long- and short-distance migrants to those found at Harworth and Ollerton were recorded. Between 1922 and 1932 58 per cent of the

[22] Interview, C. H. Green, Ollerton.
[23] Butterley L5/3 232 'Ollerton Colliery 1936–42'. [24] Ibid.
[25] Stanton Company Collieries Committee Minute 520, 8 Apr. 1929.

Table 20. *Migration to Langold 1924-32*[a]

origin of immigrants	%
Yorkshire	49.5
Nottinghamshire	26
Derbyshire	13.5
North East	4.5
Lancashire	2
Staffordshire	1.5
Wales	1
others	2.5

[a] Notts. CRO SA 105/1/1, Langold Temporary.

miners coming to the village had previously lived in Nottinghamshire; 16 per cent were from Derbyshire, and 21 per cent from elsewhere.[26] By the 1940s the number of men migrating from distant fields had declined. Between 1941 and 1948, 827 of the 1,083 men singing on at Blidworth Colliery came directly from another Nottinghamshire pit.[27] H. Tuck of Bilsthorpe states that only young men were sought by the coal company: 'Nobody was allowed to get a job if they was over 40, that was another condition. Mind you, there was a lot that was over 40 and said they were under, purposely to get a job.' The youthfulness of the workforce at the new pits is confirmed by colliery records. On 25 January 1927 there was only one man aged over sixty at Ollerton Colliery, and none over seventy, compared with 131 men over sixty at Butterley's Kirkby pit.[28] The average age of the 1,053 men who signed on at Ollerton Colliery in 1927 was just over thirty, and 79 per cent of these men were under forty years of age.[29] As seen in Table 19, only three of the 763 men signing on at Harworth in 1927 were over fifty. Just as in the case of the Welshmen who migrated to Oxford in the 1930s,[30]

[26] Notts. CRO SA 17 2/1, 2/2, Blidworth Wesleyan.
[27] Blidworth Colliery signing-on books, 1941-8.
[28] Butterley T1/6 345, 'Notts Industrial Union'.
[29] Ollerton Colliery signing-on books, 1927.
[30] Goronwy Daniel, 'Labour Migration and Age Composition', *Sociological*

young unmarried men with few ties and commitments formed a high proportion of the new population. The resulting age structure meant that within a few years there was a 'baby boom' which strained the education system and led to a very high ratio of residents per house in the new colliery villages. In 1941, 4,130 people were living in the 858 tenanted company houses in Ollerton.[31]

The 1951 census also gives evidence of unusual demographic developments in the new coalfields:

> the highest rates of natural increase occur in those districts into which a young population has been migrating in consequence of new industrial or suburban development. Worksop Rural District has the highest rate in the county (24.6%), arising from the influx of miners to the developing collieries at Harworth, Firbeck and Manton. Although the average rate in Southwell Rural District (15.4%) is lower, there can be little doubt that most of the natural increase is concentrated into the western half of the district where a new coalmining economy is being superimposed upon the older agricultural and forest landscape (at Clipstone, Blidworth, Edwinstowe, Ollerton and Bilsthorpe).[32]

However, although the *natural* rate of increase (excess of births over deaths) was high in the new coalfield between the censuses of 1931 and 1951, Powell goes on to point out that 'the general pattern of *migration* movements is almost the reverse of the natural increase . . . most of these coalfield districts also have high rates of emigration which commonly leave only a relatively small net increase of population since 1931.'[33] This was probably due both to the failure of the new coalfield to expand further after 1930 and to dissatisfaction with life in the new mining villages:

> In spite of the building of 'model villages' near many of the new collieries, the resulting communities of 3-5,000 population do not appear to be wholly satisfactory. Each of the new colliery parishes of Ollerton, Clipstone and Rufford has lost population since 1931. The situation cannot be adequately explained by the emigration of the large numbers of children who must inevitably have formed a high proportion of the

Review, 30 (1939), 281-308; *idem*, 'Labour Migration and Fertility', *Sociological Review*, 31 (Oct. 1939), 370-400.

[31] Butterley H3 224, M. F. M. Wright memo to Mines Dept. 17 Jan. 1941.

[32] A. G. Powell, 'The 1951 Census, an Analysis of Population Changes in Nottinghamshire', *East Midlands Geographer*, 2 (1954), 30.

[33] Ibid., p. 31.

young population which had settled in the villages only in the middle twenties. In the post-war period the East Midland miner, supported by his improved economic status, shows signs of preferring to live in or near the town with its greater opportunities for shopping and entertainment and for the education and employment of his family.[34]

IV

Having built up a picture of the type of men migrating to the Dukeries coalfield in the inter-war period, we must now ask why men came in such numbers to a previously rural area of Nottinghamshire, what effects such a cosmopolitan influx had, and why so many of them left so quickly. The first of these questions is certainly the easiest to answer. For a start, compared with the more unproductive declining coalfields of Derbyshire or Lancashire or Durham the new field could offer the secure knowledge that its pits would enjoy many years of life, perhaps running into centuries: the Dukeries had 'great expectations', certainly greater than those of older mines struggling in marginal profitability.[35]

Closely connected with this, the new pits worked fuller time than the old, which meant more take-home pay for the miners. Herbert Tuck's father had been working at Silverhill Colliery, Sutton-in-Ashfield, West Nottinghamshire: 'You could go a full week and get one day in and no dole money,[36] and this is how it went. Anyway my father was in this system, kind of thing, and when he heard about this new pit being sunk at Bilsthorpe . . . he could foresee that once the shafts were down there was going to be job security.'[37] *The New Statesman* summed up the financial reward for miners in the new coalfield in 1927:

Wages are high in the Dukeries' pits. No black shadow of the 'nine-hour minimum' haunts the labourer and day-wage men here, as it haunts the rest of the colliery industry. In the Bolsover pits some coal hewers are drawing up to a pound a shift; ten miles away in Leicestershire it drops to 10s., twenty-five miles away in South Derbyshire it sinks even beneath that miserable amount. Time is good; 'short time' is

[34] Ibid., p. 38.

[35] See below, ch. 3, for the 'great expectations'.

[36] This was because they were made to turn up at the pit even if there were only one or two hours' work each day, rather than concentrating the work so that they could draw dole money on the days on which they were 'unemployed'.

[37] Interview, H. Tuck, Bilsthorpe.

abnormal. In the last twelve months, when the rest of the Notts field has thought a four-day week a God-sent gift, the new pits have been turning five and six, and even seven shifts a week.[38]

It is scarcely surprising that in July 1929 the *Newark Advertiser* could report that

Each day more and more miners apply for work at Ollerton Colliery. Owing to the pit at Barlborough being flooded, many of the men have signed on at Ollerton. There is a large queue every day of men from Worksop, Retford, Doncaster, and Mansfield. The pit is on full time, and the output is increasing. 2,500 tons per day are being raised. This is one of the few pits where full time has been worked all the year. Houses are still being built. There is nearly another 100 completed.[39]

In fact the new pits did not all work full-time, especially when the quota system was in operation during the 1930s; but they did usually work more shifts per year than the older collieries. Between 1928 and 1936 Clipstone Colliery always put in between 197 (1931) and 243 days (1935); another Bolsover pit, Thoresby at Edwinstowe, worked between 203¾ (1936) and 274¼ (1930). These figures were consistently higher than those for other Bolsover pits. In 1936, for example, Clipstone worked 220¾ days and Thoresby 203¾; only Creswell (202) of the Bolsover Company's other mines worked more than 200 days.[40] Similarly the 227¼ days worked by Ollerton Colliery in 1938 was the highest recorded of all the Butterley pits, just as it achieved the highest output tonnage, highest output per face-worker, and highest output per employee.[41]

In general we can say that the miners who came to the Dukeries field in the 1920s and 1930s from older pits would be classed as 'resultant' migrants in the terms of A. H. Hobbs;[42] that is to say, if it had not been for the threat of redundancy and the poor wages they received in the old pits they would not have moved. Few aspired to any dramatic improvement in their position, and if any of the migrants shared the high hopes for the prosperity of the new coalfield it was only in

[38] *New Statesman*, 24 Dec. 1927. [39] *Newark Advertiser*, 17 July 1929.

[40] Bolsover Company Annual Reports, 1928–36.

[41] Butterley A4/14 17, 'Colliery Development'.

[42] A. H. Hobbs, 'Specificity and Selective Migration', *American Sociological Review* (1942), 772–81; R. C. Taylor, 'Migration and Motivation' in J. A. Jackson (ed.), *Migration*, pp. 116–22.

contrast to the absolute decline of their home districts. There is evidence that resultant migrants are more likely to find adjustment to their new environment difficult than those 'dissenting' migrants who see a move as a positive opportunity.[43]

V

What were the effects of such a considerable influx of men and their families into—and through—the new Notts. coalfield in the inter-war years? Specific consequences such as those concerning education,[44] trade unionism,[45] local government and party politics[46] will be discussed elsewhere. However it is clear that in general there was a degree of chaos in the new mining villages in their early days:

RJW You say that people came from all over the country to live in Bilsthorpe. Did this mean that in the early years of the village it seemed very cosmopolitan?
HT Oh, proper League of Nations!
RJW But did these different groups of people get on well together?
HT No, I would say that when the village first started there were hell let loose which normally even to this day happens. When you get communities from all over the place drawn together, because whereby you might get eighty good families, you always get twenty wanderers. You know, the outcast kind of thing if you want to put it that way; and these people, a new village, a colliery village, it takes time to settle down. It'd maybe take ten years, but within that ten years, those—the wanderers—leave and go somewhere else, and you'd get the other ones and eventually of course you do get a stable population.[47]

Christopher Storm-Clark described the contrast between the unstable population and the confusion of coal-mining communities such as those in the new Dukeries field and the traditional, well-established pits where sons followed in the footsteps of their fathers to the coal face:

A 'family' pit means a well-organised pit around which there revolves a stable mining community. The opposite type of pit is called 'cosmopolitan', so that cosmopolitan pits are to be found in, for example, the

[43] R. C. Taylor, 'The Implications of Migration from the Durham Coalfield: an Anthropological Study' (Durham Ph.D. 1966), p. 230.
[44] See below, ch. 7.
[45] See below, ch. 5.
[46] See below, ch. 6.
[47] Interview, H. Tuck, Bilsthorpe.

prosperous and expanding North Nottinghamshire coalfield, where immigrants from Scotland, Wales and Durham make up for a deficient local labour force.[48]

The differing origins and characteristics of the men coming into the new villages were stressed at length in one of the few pieces of contemporary description of the coalfield, an article in *The Sphere* in 1932 which wrote of the colliery village of New Ollerton:

I did not need to go 550 yards down to discover that the best miners are Derbyshire men, usually from Ilkeston, Heanor or Eastwood,[49] though South Normanton men claim they are just as good if not better . . . at Ollerton they always like to get Heanor men. Occasionally they will send over a couple who hailed from there originally for a weekend, with all expenses paid, to induce more of their mates to join them.[50]

Derbyshire men could be identified by their moleskins changed for 'knickers' at the pit, their dialect, and their habit of carrying their water-bottle in their pocket and their snap-bag over their arm. The Welshmen, 'who are as small as the Derbyshire men are tall, are sharp-featured and hot-headed.' They always carried their water-bottles attached to a rope over their backs, and wore clogs: 'There are less than a dozen of them at Ollerton. Those that remain are good men. But no more are wanted. As a race they talk too much politics and cause too much trouble, and they are not popular with the other miners.' The Durham man carried his bottle in his hand, turned up late for his shift, and his trousers reached only three-quarters of the way to his feet. The Lancashire man wore clogs and carried his water in a Dudley bottle on his back. The Staffordshire man 'with his sing-song dialect . . . is a very good type of miner, but he can never make up his mind whether to carry his water bottle in his hand or on his back.' The Yorkshire man liked 'easy' coal and grumbled at using a pick.

The Irish are magnificent sinkers—the men who sink the original shaft, work which calls for courage and immense strength. They have glorious

[48] C. Storm-Clark, 'The Miners 1870-1970, a Test Case for Oral History', *Victorian Studies*, 15 (1971-2), 67.

[49] Actually in Nottinghamshire.

[50] Charles Graves, 'A Coal Miner's Life Above Ground', *The Sphere*, 23 Apr. 1932, pp. 140-1, 162.

thirsts. At a dinner given by the colliery to celebrate the conclusions of the sinking operations, 180 of them drank 216 gallons of beer, which is a little over nine pints a man—and the local beer is dark and strong. They are men with whom you should never discuss religion or Ireland.

'The Scots, I am told, are just plain, bigoted, rough fellows.'[51]

This catalogue of caricature and generalization does in fact represent a very important feature of the early stages of the new colliery villages—an obsession with regional differences, distinctions between what *The Sphere* called 'races'. It is scarcely surprising that men came to identify more with their own regional group than with their workmates as a whole. The *New Statesman* stressed that the region had no mining tradition of its own:

It is a new coalfield with no mining population upon which to grow. There is no mining tradition. The existing local ideas were based on centuries of agricultural feudalism, which looked towards the squirearchy and the county nobility, which was centred on the great houses which dot every part of the Dukeries. The small towns—Ollerton and Bawtry—are only today awakening from the facts of medievalism. In this closed area the miner was an intruder, an eyesore and an anachronism. From every dying coalfield, from every rich coalfield, from far and near, the miner has flocked into this new paradise. He has come from the poverty-stricken villages of Abertillery, and Blaina and Ebbw Vale, from tiny South Derbyshire, from neighbouring Leicestershire, from the hot hells which are miscalled pits in South Yorkshire, from the older pits in the Leen and Erewash Valleys on the borders of Derby and Notts, from that place of desolation and starvation that used to be proud six-hour day Durham. The new mining population is heterogeneous from top to bottom.

Local isolation dies hard. The men bring with them their old customs and their old local outlook. To the Durham man and the South Wales man the Dukeries is still a passing haven. The melting pot is working slowly; no unity of feeling has yet been created. There is a Dukeries coalfield; there is as yet no Dukeries miner.[52]

There is evidence that the coal companies sometimes exploited the regional differences amongst the miners, possibly to prevent the establishment of the feeling of a workforce united against the management, or to provide scapegoats in a similar way to the manner mining companies in Africa would use tribal differences.[53]

[51] Ibid.
[52] *New Statesman.*
[53] See for example Charles Van Onselen, *Chibaro* (London, 1976), chs. 5-6.

RJW Did you find that any other[54] sorts or groups of people tended to congregate in another part of the village?
GC Well, a little bit, yes. There was a lot come down from Durham, and they appeared to hang about Fifth Avenue.
FT That was later on, George, when they were the refugees from the North East.
GC That was in the thirties, early thirties.
RJW And how did they come to stay in one place, could they choose which house they went into? Or were they just given a house?
GC No, not as such. We were handed six keys, to take our pick, but I don't think everyone was.
RJW Well, how did the Geordies get together then?
GC Well, that was the Cashier, Russon, he was the man that told you where you were going to live.
RJW So the company actually intended to put people in groups like that?
FT Yes.
GC Oh yes. I think they grouped them up the Fifth Avenue.
FT It was class distinction. Nothing more, nothing less, from Russon's point of view. He wanted all the scruffs, scabs, or whatever you want to call them in one place, out of the way, as far away from him as he could get them. He did, and they were right down the bottom of Fifth Avenue, weren't they!
RJW What did people think of the Geordies? Did they think of them as rough?
FT Lunatics!
RJW Why, what did they get up to?
FT I mean, every verse was a Geordie joke, wasn't it? It didn't matter what happened, it was a Geordie joke. There were some good men among them, but a lot of them went back, they didn't stay. Work was too great. I mean I heard several yarns from underground where one bloke was doing one job and he expected somebody to pass him part of what he was doing. Well that wasn't on at Thoresby, was it?
GC Never.
FT You did two jobs, not one. That's why there's so few Geordies left compared with the number that came down.[55]

In the above extract we see the role of the company in deliberately isolating a regional group in the village, the idea that the Geordies were a separate 'class', their role as the butt of jokes, and the idea that they were inferior workmen. This view of the division within the colliery village is corroborated by another interviewee who lived in Edwinstowe:

[54] i.e. besides officials grouped in deputies' rows.
[55] Interview, George Cocker, Frank Tyndall, Edwinstowe.

Well this was Fifth Avenue in Edwinstowe just the same. That was the dirty avenue of Edwinstowe, and soft as it sounds, all the dirty families *were* on it in those days . . . it was funny how, some nice people amongst them, but in the main the bottom part of Fifth Avenue from the junction of Fourth where the school was, they were all dirty families and really big families, and the rest of the village everyone seemed so nice and respectable.[56]

Regional origins were never forgotten in the new Dukeries colliery villages. As late as July 1938, the local newspaper reported under the heading of 'County Cricket at Ollerton' that a cricket match had been played on the Walesby Lane ground between colliery employees born in Nottinghamshire and those born in Derbyshire, as part of the Ollerton holiday week programme.[57] To whatever extent these divisions were fostered by employers, they had remained an integral feature of the mining communities.

VI

Why did so many of the new arrivals in the Dukeries coalfield depart so quickly? It is much more difficult to judge why men left the new villages than why they came to the Dukeries: one can, of course, not find many miners who only lived in the area for a brief period in the 1920s. Even allowing for the expected disappearance of the sinkers who specialized in a particular kind of work, the figures for labour turnover show that the workforce remained in a state of flux for at least ten years after the pits had reached productive status. It has been said that men from distant coalfields found the working conditions unattractive.[58] Yet long-distance migrants never made up more than a quarter of the immigrants in any of the new pits, and the statistical evidence available does not seem to indicate that they were any more likely to leave than short-distance migrants, perhaps because of the greater commitment involved in moving a long way.

Perhaps all those who came from different working conditions found in older collieries found it hard to adjust to the deep pits of the new coalfield. Neville Hawkins came to

[56] Interview, Barbara Buxton, Forest Town (Edwinstowe).
[57] *Ollerton, Edwinstowe and Bilsthorpe Times*, 15 July 1938.
[58] See above, p. 47.

Harworth from Grassmoor in North East Derbyshire, only about twenty miles away, in 1931:

NH When I came here, I came in the stream of it, and the pit village was only six, seven years old then, and people were coming and going every day.

RJW A high turnover of labour?

NH Yes. Many men were from, like myself, the shallow mines, the very shallow mines, which in the main are cool, cold mines in which to work, and found the pressure of a deep pit like Harworth far too much for them, and they were coming and going, and they had a week at it and packed up and flitted back to where they had come from. They couldn't stick it.[59]

Most pits sunk in the nineteenth century were under 500 yards deep. Ollerton reached 650 yards, Blidworth 742 yards, Thoresby 756 yards, Bilsthorpe 792 yards, Harworth 978 yards, and Clipstone 1,006 yards.[60]

However, the warm working conditions were not the only reason for the high turnover of labour in the new coalfield. It is clear that the Dukeries disappointed the immigrants in a variety of ways. Services such as miners' welfares, schools, water and sewerage schemes, new roads, colliery pithead baths, and canteens were very slow to arrive. The rawness and isolation of these mining villages planted in a hitherto rural area meant that in the early days a ten-mile bus-ride into Mansfield or Worksop was necessary for special entertainment or weekly shopping. Meanwhile, except at Bilsthorpe, the newness of the pits was no guarantee of full time, and though three days a week was better than that achieved in many other areas, the economic attractions of the Dukeries seem for many not to have outweighed its deficiencies. Nor was job security as great as at first imagined; the coalfield did not develop to the extent which had been expected during the unfavourable trading conditions of the 1930s, and the companies were on occasion forced to lay off as many as 300 men at a time.[61] At Blidworth, an unexpected fault discovered in 1929 led to the temporary closure of the colliery, and the laying off of all its miners. Nevertheless, it is possible to argue

[59] Interview, Neville Hawkins, Harworth.

[60] NCB, *Coal in North Notts*, 1977–8.

[61] See above, Table 12, Aug. 1932.

that the main causes of discontent in the new coalfield were social and political rather than economic.

Many miners missed the atmosphere and facilities of the established colliery villages from which they had migrated. In particular they maintained links and ties with the extended families they had been forced to leave behind:

AEC They certainly kept strong links with their area. For instance they didn't always go to the most convenient shopping place, they would go to Mansfield or further afield to Sutton quite often where they would meet more and more of their old friends on market days.

RJW And they took their old local newspapers and so on?

AEC Yes, that was very noticeable, and most of the houses you went in, the *Derbyshire Times* of whatever district they had come from was always in evidence.[62]

Sometimes this homesickness could actually lead to departure from the village:

One of the things that I noticed when I was teaching at the Dukeries, Ollerton, when we had the influx from the North East when they closed the pits down in the Durham coalfield, was that a number of those didn't stick it. They were kids who you'd had for one term and then they'd be gone back. They'd obviously missed the social life, or the surroundings, or whatever.[63]

The new coalfields, at least in their early years, did not conform to the classic picture of the 'neighbourliness' of mining communities. In Kent, the practice of gossiping in one another's houses, common on the older coalfields, was missed by the women who had migrated with their husbands.[64] The manager of Ollerton Colliery concluded in 1939 that: 'I am rather afraid that colliers are hardly the class of people to be tenants of flats. They work on various shifts, they live in varying degrees of cleanliness, and also do not appear to be generally friendly with their neighbours.'[65]

As dealt with more fully below,[66] the coal companies maintained a considerable degree of control over the lives of their employees outside the pit as well as at work. All the

[62] Interview, A. E. Corke, Ollerton.
[63] Interview, Cyril Buxton, Forest Town (Ollerton).
[64] See below, ch. 11.
[65] Butterley H3 224, W. S. Fletcher, 20 Dec. 1939.
[66] See below, ch. 4.

housing was company-owned, and tenancy was subject to a large number of regulations: rarely were pets or animals of any kind allowed, gardens had to be kept neat and tidy, and noisy or 'immoral' behaviour was frowned upon. Militant activity in trade unionism or politics was not tolerated, and the naturally divisive butty system, unknown in many other coalfields, was prevalent in the Dukeries. This stern discipline and domination was not to the taste of all miners, and, denied the opportunity of any other form of protest, many left the village voluntarily if they were not evicted:[67]

> There was another point also which we mustn't forget, that the discipline in the pits was really slavery, and in those days if you had a bit of spirit, kind of thing—and maybe this is where the wanderers came in[68] —if you had a lot of spirit and said well, look, you know, you don't think this is fair, and I even experienced that in my early days and I got the sack for six weeks, even as a youth, because I'd got a lot of spirit. There was a great movement of labour that road. If your face didn't fit, you were out! Now they could throw you out of your house overnight, and this is what was happening all the while, kind of thing, as I say until they got this settled population.[69]

As in many other paternalistic communities where power was concentrated in the hands of a single authority such as the employer in a factory village or single-industry town, the twin responses to powerlessness on the part of the workers were a deferential acquiescence in the 'beneficent' control of the management and an impotent resentment. This could be crystallized in the attitudes of the miners to the figure of the colliery managers, who tended to be regarded either with great respect or else with considerable concealed bitterness. One resident of Harworth had no high opinion of the colliery manager and undermanager: 'All I know, Wright just run this pit and he was Him, he were the king. Do you understand me? The Gestapo. Him and a fellow called Pedley, down the pit, he were undermanager, another Gestapo. Honestly he were the rottenest man I've ever met in my life!'[70]

[67] For example the cases in the Butterley records of Dowdall, evicted for refusing to part with his greyhound (see p. 100) and Sam Booth, evicted for trade union activities (see p. 124).

[68] See above, p. 44.

[69] Interview, H. Tuck, Bilsthorpe.

[70] Interview, Tommy Jenkins, Harworth.

Just as it was often immigrants from more distant coalfields who formed the basis of political opposition to the coal companies in the 1940s,[71] men from outside Nottinghamshire apparently resented the control of the management in the community more than their colleagues:

But those that came in, from Durham and Yorkshire, Yorkshire in particular, they were up in arms, they'd never been used to it, and so they weren't going to accept it, they'd been used to more freedom. And so underground it was a constant coming and going, like every day there was men set on here. There were hundreds of men.[72]

Another factor which led men to leave the Dukeries was the lack of employment for women and girls, which meant that the family income could not be supplemented in the way that it could in, for example, western Nottinghamshire, where the hosiery and textile industries had long been established in coalfield towns like Sutton-in-Ashfield and Hucknall. This was a problem recognized by the Butterley Company, who built a factory in Ollerton in the later 1930s, to provide employment for the wives and daughters of their miners.[73]

These young and isolated colliery villages could offer little entertainment of the kind traditionally favoured by miners and their families. As the Stanton Company's Colliery Committee put it in August 1929:

You will notice from the details given in this report that a considerable number of men are set on during the month, and that a considerable number of men also leave. This is natural at a new Pit, but in the case of Bilsthorpe I think it is rather abnormal. The opening of a Cinema and the provision of a Barber's Shop etc is having attention and these will no doubt improve the amenities of the Village. Bilsthorpe is a considerable distance away from any town, where there are any shops etc.[74]

There are, then, many reasons why miners left the new Dukeries coalfield in its early years. Like many other pioneering 'settlers', the immigrants to the newly founded colliery villages found disadvantages to counterbalance the promise of economic security. Many preferred the life they had known in the older coal-mining communities of Britain, but they

[71] See below, ch. 6.
[72] Interview, Charles Stringer, Harworth.
[73] See below, ch. 4.
[74] Stanton Company Collieries Committee Minute 559, 12 Aug. 1929.

were replaced by others keen to take part in the industrialization of the Dukeries. This accounts for much of the turbulent character of the new coalfield. Moreover, the dislocation and disruption which occurred in the new Nottinghamshire field of the 1920s and 1930s should not be regarded as unique, although the economic conditions of those years and the isolation of the Dukeries in an age before mass transport and mass communications clearly strengthened the hand of the coal companies and increased the difficulties of adjustment to new surroundings. Not only are there parallels with what we know of the early stages of other coalfields in Britain in the nineteenth and twentieth centuries,[75] but more recent colliery openings have provoked similar problems. Cotgrave in South Nottinghamshire, where a pit was opened as late as 1964, became a 'problem area' of rampant crime, vandalism, alcoholism, and marital strife despite its relatively high 'standard of living'—and Cotgrave too suffered from a notably high rate of labour turnover.[76] The social and economic problems posed by the suggested Belvoir coalfield have led to full-scale studies by the NCB and the Institute of Planning Studies at Nottingham University. Sociologists have often studied the static elements of coal-mining communities; perhaps more attention ought to be paid to their dynamic nature. Often, under close scrutiny, stability proves to be illusory.

[75] See below, ch. 11. [76] Ibid.

3. Reactions

'A little replica of the Industrial Revolution'[1]

How did contemporaries view the economic and social transformation of the Dukeries region of Nottinghamshire when seven new pits were sunk in the 1920s in what had previously been the rural seat of a number of great aristocratic landowners, an area in which the remains of Sherwood Forest were divided between the estates of the Duke of Newcastle, the Duke of Portland, Earl Manvers, and Lord Savile? Did the inhabitants of the old villages of Edwinstowe, Clipstone, Ollerton, Bilsthorpe, and Blidworth fear and resent the 'invasion' of immigrant miners, or did they welcome the wealth and trade that the pits might bring? Did the Dukeries' aristocrats fear that their influence within the region, based upon a pattern of landownership and deference built up over many decades, would be weakened by the arrival of the coal-owners and their employees? Or did they regard the opportunity to lease out part of their estates and claim royalties from every ton of coal dug from beneath their land as a last chance to maintain their country retreats and the standard of living to which they were accustomed? How did the aggressive and efficient coal companies who were determined to exploit the new coalfield get on with the old élites in the villages in which they took so much interest in order to protect their investment? And how did the miners themselves get on with the existing residents of their new surroundings?

We must first consider the nature of the district concerned before mining arrived in the years after the First World War. In most of the villages, there was no manufacturing industry of any kind; the only sizeable sources of employment were agriculture and the estates of Lord Savile at Rufford Abbey, Earl Manvers at Thoresby Hall, the Duke of Portland at

[1] C. Day Lewis, *The Buried Day* (London, 1960), p. 131.

Welbeck Abbey, and the Duke of Newcastle at Clumber House. In the nineteenth century, Nottinghamshire had ranked third amongst English counties in order of the proportion of the total area occupied by estates which exceeded 10,000 acres; its figure of 38 per cent was exceeded only by Rutland and Northumberland.[2] In the 1920s, the pattern of landownership was little changed from this. However, it would be wrong to see this region as nothing but a backwater, for it lay in the heart of Sherwood Forest, by now much cleared but still arguably 'the prettiest stretch of country north of Sussex and south of the Highlands'.[3] There was a flourishing tourist trade in the summer, particularly centred on the village of Edwinstowe, near the famous Major Oak. An eighty-six-year-old resident of Edwinstowe remembered the popularity of the village:

RJW Have you lived here all your life?

HP Aye. You want to know about it? Quite peaceful little village, invaded with thousands of trippers in the summertime, from Lancashire, Yorkshire, and everywhere; it were a real centre for people coming to visit the Dukeries, all owned by lords and dukes, that's why it's called the Dukeries. It's many many years since, before the First World War . . . they used to come—oh, I've seen streams half a mile long, horses and brakes and that from Worksop.[4]

Nevertheless, the pace of change in these old villages, the largest of which, Edwinstowe, had a population of under a thousand in 1921, was dramatically transformed by the sinking of the mines and the building of model colliery villages for the miners.

It was recognized at the time that the economic change taking place in the Dukeries in the 1920s and 1930s was a very significant one. The new coalfield was seen as a great new venture offering lucrative rewards for the industrialist prepared to make the heavy investment required to sink a deep pit and build a new village in this isolated area. As the

[2] J. Bateman, *The Great Landowners of Gt. Britain and Ireland* (1883), quoted in F. M. L. Thompson, *English Landed Society in the 19th Century* (London, 1963), p. 32.

[3] *The Sphere*, 23 Apr. 1932.

[4] Interview, Harry Parnell, Edwinstowe.

New Statesman wrote on the striking of the Barnsley bed of coal at Bilsthorpe Colliery in 1927:

Bilsthorpe is the latest, but far from the last, of the long line of pits which in the last twenty years have transformed the glades and villages of Sherwood Forest, and the Dukeries, into one of the greatest coalfields in the whole of Great Britain. Steadily, quietly, till we stumble on their existence with a shocked surprise, pit after pit has been sunk and worked in the beautiful stretch of country which covers the eastern half of the county of Nottingham. It is one of the romances and revolutions of modern industrialism. As the older pits on the coal seam which outcrops on the borders of Notts and Derbys have shown signs of exhaustion, the coalmasters have consistently followed the seams eastwards. They bored and proved coal in the whole of East Notts up to the Trent and beyond the Trent, till they found it 4,000 feet under Lincoln Cathedral. Twenty years ago the marvellously rich bed of the Barnsley seam was tapped in South Yorkshire; a dozen pits whose names are almost household words—Maltby and Dinnington and Rossington and Bullcroft—were dug to exploit it. It was the stimulus needed; the 'scramble for the Dukeries' followed fast. The giants from South Yorkshire, Lord Aberconway and Charlie Markham, came down to open Blidworth pit in the south. The Bolsover Company flooded out from Mansfield in an astonishing expansion. The local coalmasters—the Wrights of Butterley, the Seelys of Babbington, the Barbers and Walkers —were joined with those who scented profit from distant Lancashire and more distant Scotland; all came to reap the golden harvest of the Dukeries.[5]

This entertainingly overwritten piece is worth quoting at length because it captures the 'gold rush' spirit of the 1920s in the Dukeries. There was a feeling that this great new field could rescue the coal industry from decline and offer an alternative to the small, unproductive operations now closing throughout the older coalfields of Britain:

In every material respect they have made and are making the Dukeries a great coalfield. The pits are all new, all on a vast scale. Here there is no decrepit striving after 500 or 600 tons a day. The Dukeries thinks in thousands of men and millions of tons. Each of the five pits of the Bolsover Colliery Co., which towers above all rivalry, employs nearly 2,000 men or more; their joint output is now 4 million tons a year. Three thousand work at Manton, even more at Warsop Main. The collieries that are developing at Edwinstowe and Ollerton and Thoresby[6] under the shadows of the famous houses that ruled the Dukeries, will

[5] *New Statesman*, 24 Dec. 1927.

[6] Thoresby Colliery was situated at Edwinstowe, so only two different pits are referred to here.

reach or surpass the same level. Their coal is better than the best South Welsh. Their seams run to six feet and six feet six; there is no room here for the miserable two foot nine and three foot scrapings of other fields.[7]

The coalfields did not live up to some of the more extravagant claims made for it. A. J. Bennett, the Liberal MP for Mansfield, said that 'perhaps in four or five years time there would be something like 80,000 more people in this district' at a meeting of the Nottinghamshire Church Extension Society in Mansfield Town Hall on 23 April 1923.[8] In fact the colliery companies' plans were cut back as it became clear that conditions in the coal trade would not be as favourable as they had been before the Great War. Butterley originally expected to build 1,200 houses in Ollerton; they completed 832. Early hopes for the prosperity of the coalfield also led to overambition at Bilsthorpe. When the wooden church of St. Luke's had to be closed in 1939, the local paper wrote that:

When St. Luke's was originally built it was understood that Bilsthorpe was to grow into a large village of 5,000 or more inhabitants and that the present wooden church would become a church hall when a new brick church was built. A site for this new church was chosen and plans drawn. These hopes were not realised, however.[9]

The population of Bilsthorpe was still less than 3,000, and it was very unusual to have two churches in one parish—after the construction of St. Luke's there was a church in the colliery village and one in the old village of Bilsthorpe. As there was no division of the ecclesiastical parish, 'it did result in the splitting of the parish into two "camps"'.[10] It was not only the religious authorities that were misled. In the autumn of 1925 the education authorities were still working on the assumption that 1,560 houses would be built to serve Bilsthorpe Colliery.[11] In the end, the Stanton Company built little more than 400.

The village of Edwinstowe was not completed by the Bolsover Company either. In April 1924 the company announced plans for the building of a colliery village of

[7] *New Statesman.* [8] *Mansfield Reporter*, 27 Apr. 1923.
[9] *Ollerton, Edwinstowe and Bilsthorpe Times*, 2 June 1939.
[10] Ibid. See below, p. 183 for the quarrel between the manager of the colliery and the Rector of Bilsthorpe.
[11] PRO ED 21/37496 Bilsthorpe Council School, 19 Sept. 1925.

956 houses.[12] Eventually the model village consisted of 497 houses:

You see originally this is half the village, really. The Coal Board own all the land from here right to the Archway, that's the first woodland up the road there. So the tennis courts, bowling green, club, welfare, clinic, and school were actually the centre of the village. But work progresseed at Thoresby, and they didn't want the number of men in the end that they thought they would do, so that actually the cricket ground, now that should have been the centre of the village attraction, is the outside of the village, because the village has never been extended.[13]

But all the same, even if the Dukeries field did not quite live up to the wildest expectations, *beliefs* about its future expansion are important when considering contemporary attitudes to social and economic change; and after all, production and employment in the Dukeries *did* boom while the traditional bases of British mining declined after 1918, and the relatively new pits of Nottinghamshire and South Yorkshire *did* become the mainstay of the nationalized coal industry.

The effects of this revolution were perceived in a number of ways at the time. The clearing of the forest when the pits were sunk was ripe for symbolic interpretations of the process of industrialization. A collier's rhyme in Ollerton ran:

When this pit was sunk in 1923
On this spot stood a great oak tree.
Ever since they made that great big hole
The only cry is 'keep on sending up more coal'.[14]

This verse would not have impressed the poet C. Day Lewis, who lived through this period at Edwinstowe, where his father was vicar. Day Lewis reported his own memories of this time in his autobiography:

When my father moved to Edwinstowe, it was a country village. Before he died, it had become a mining town . . . We lived on coal. Seams of it lay below our feet–rich seams which had hardly been tapped yet, for the north Notts coalfield was the youngest in the country; and my father's stipend of £600 a year came largely, I believe, from the titled patron of the living, beneath whose land the coal had been found. Had I

[12] *Mansfield Reporter and Sutton Times*, 18 Apr. 1924.

[13] Interview, F. Tyndall, Edwinstowe.

[14] R. H. Mottram and Colin Coote, *Through Five Generations* (London, 1950), p. 118.

possessed any historical sense, I could have observed during our years at Edwinstowe a little replica of the Industrial Revolution being constructed before my eyes. When we first went there, the black-faced miners cycling home from their work with their snap-tins bumping at their sides were outnumbered by the farm labourers whom they passed in the village street, exchanging a curt Midlands 'how-do?' Twenty years later, almost every man would be working in the pits, just as in the course of those years the feudalism which had survived through the great landed estates about us gave way to the more anonymous but on the whole fairly enlightened autocracy of the big colliery companies: the blacksmith's forge had become a garage, and new housing estates obliterated the contours of the fields which as a boy I had seen tossing with oats or barley. . . . In more ways than one I lived in Edwinstowe between two worlds.[15]

The local newspapers made more prosaic comments on the transformation, which they noticed particularly when the growing population of the Dukeries necessitated the announcement or the completion of a local government water, sewerage, or housing subsidy scheme. The *Worksop Guardian* reported on 7 September 1923 that

there was an avalanche of plans at the meeting of the Blyth and Cuckney Rural District Council held at Worksop on Wednesday. The Council Chamber was filled with rough sketches, plans and drawings, which were piled up on top of each other, and occasionally it was impossible to see the members of the council!

Similarly in 1926 the *Newark Advertiser* wrote of the prospects for the Southwell Rural District Council area:

Coal spelt with a capital C is rapidly changing the aspect of the Southwell Rural District area from a pastoral to a mining character. Those who travel by motor car through the Dukeries must be impressed by the number of new villages for colliers now being built. The Southwell RDC meetings indicate also the health, housing and educational provision which are going on there. It is estimated that by 1930 the amended output of the new pits of the county will be 20 million tons, or more than half the production of Scotland, and will last until the year 2,300 A.D.[16]

B. S. Townroe wrote in the *Spectator* in 1926:

Those who have travelled at Easter-time by car through the 'Dukeries' in Nottinghamshire must be impressed by the number of new villages for colliers now being built. The development of a great new coalfield

[15] C. Day Lewis, *The Buried Day* (London, 1960), pp. 130–1.
[16] 'Changing Southwell', *Newark Advertiser*, 14 Apr. 1926.

of South Yorkshire and Nottinghamshire, covering an area of not less than 600 square miles that contain rich seams of good quality coal, is causing a migration of populations in the Midlands . . . the provision of accommodation for the miners and their families presents a series of problems for those directing the new enterprises. In the houses that I have visited recently, the standard of amenities provided is equal to that of any housing scheme in the country. The Coal Commissions' Report in this matter states: 'With regard to the new collieries, we observe with great satisfaction the efforts that are being made in a number of instances to provide a sufficiency of accommodation, and of a proper standard.' At some of the large new mines on the coalfield, from one third to nearly one half of the total capital provided is being spent on housing.[17]

On 17 August 1927 the Union Jack was hoisted at Bilsthorpe as coal was reached after two and a half years of sinking work. The *Newark Advertiser* reflected as follows:

Since sinking began the rural peace in which old-time Bilsthorpe slumbered has been rudely shattered. The colliery chimneys now dominate the landscape, while in addition to the 'bungalow town' in which the sinkers made their homes, the nucleus of a permanent model village is already in being. Over 200 houses are already occupied, and it is expected that the number will eventually reach 1,000.[18]

In December 1927 the same paper described the 'town' of Ollerton in an article occasioned by the opening of a £20,000 sewage disposal plant and water tower. Southwell Rural District Council had only recently begun to have to undertake such work, but Chairman Alexander Straw believed that they had 'shouldered their responsibilities'.[19] When the Carlton/Hodsock sewage scheme was opened in 1926, similar speeches on the theme of 'progress' were precipitated. A. Brunyee, the Manager of the Midland Bank in Worksop said that a few years earlier the Rural District 'seemed to be but a small rural affair, but time and progress, and particularly the coal developments in the area, had brought it into the foreground and into a very responsible position.'[20] Two years later the *Worksop Guardian* reported the ceremonial opening of a new water scheme for the same district: 'Until the development of the two new collieries in the area–Firbeck Main and Harworth –the district was purely rural and the population small and stationary.'[21] However, continuity was provided by the Rural

[17] B. S. Townroe, 'A Miner's Village', *The Spectator*, 10 Apr. 1926.
[18] *Newark Advertiser*, 24 Aug. 1927.
[19] Ibid., 7 Dec. 1927.
[20] *Worksop Guardian*, 3 Sept. 1926.
[21] Ibid., 27 July 1928.

District Council's Chairman, eighty-five-year-old former farmer William Ghest, here described as 'not only a very good Chairman but also a perfect type of English gentleman (Applause)'.[22]

But the local newspapers did not always stress this theme of 'progress'. Occasionally hints were given that the arrival of coal in the district was not greeted with whole-hearted approval. In October 1927 the *Newark Advertiser* weighed up the pros and cons:

> A transfer of population, such as the new pits in Notts has brought about, intensifies the difficulties of one district, if it does something to relieve the troubles of others whence they came. There is the obvious retort that new properties such as coal pits and railways, new shops, stores, public houses etc. may bring more money into the district and more rateable value, but experience shows that the set-off is pretty considerable against these.[23]

The particular problem which caused these reflections was the appearance in Southwell Police Court of fifty miners unable to pay their rates in the aftermath of the long coal strike of 1926. On occasion the darker social fears unleashed by the influx of miners into this deeply conservative agricultural area were revealed, or as the *Newark Advertiser* put it, 'an influx of folk alien alike in thought and tradition, in outlook and purpose to the natives and rural folk amongst whom they have set themselves down.'[24] The hidden thoughts underlying such a view were more openly stated by the *Mansfield Reporter* of 7 March 1924, in which the following appeared under the heading 'New Colliery at Blidworth':

> It was said that probably before the builders left the village they would erect 2,000 houses. That means an enormous change for the quaint old-world village of Blidworth, with its centuries-old rocking service.[25] In the past the village has been regarded as purely agricultural, although of late years, since the coal pits have sprung up—or should it not be sprung down?—the colliery workers have invaded the village and the undefiled air of the parish has been polluted with Bolshevic talk.

For the existing residents of the Dukeries, the arrival of mining meant that rates rose to pay for unwanted council

[22] Ibid. [23] *Newark Advertiser*, 5 Oct. 1927. [24] Ibid.

[25] The ceremony of 'rocking' the last-born baby boy in a particular section of the church calendar (interview, Marjorie Wilson, Blidworth).

and county council action, schools were 'invaded' and over-crowded, and even the shopkeepers and traders hardly benefited from an increase in custom since the immigrant miners brought immigrant shopkeepers from the older coal-fields in their wake. Applications to open shops in the Butter-ley Company's colliery village at Ollerton came from Bolsover, Nottingham, Forest Town, Langwith, and Heanor.[26]

As an Ollerton resident complained in 1936: 'I understood that when this pit came here it wouldn't interfere with the old village. Now they've only to go and look around to see the empty shops and bare windows. They've built new shops up in the new village and left the old village dormant.'[27] It would seem therefore that the traders did not welcome the influx of miners; Sam Brown, the colliery checkweighman, agreed. Replying to Mr Hedge, he said:

> I'm afraid he's feeling rather sore, and the colliers will quite understand his soreness, and if the positions were reversed, we should feel just as he's feeling, and express it. It's true that the tradesmen of the old vil-lage have lost; some shops have closed, but the simple fact is, that the old shops could not possibly have catered for all those people. . . . Even though they may have been happy with their arrangements in the old days, they must admit progress.[28]

Various residents of Old Ollerton complained of the loss of 'the good old days' in *Manor to Mine*. The hedger Spider Parkin found less work and higher rents:

> Now of course, the pit's come here. I find this pit, as regards my work, has destroyed a matter of anything from 6 to 10 miles of hedging, so that doesn't want doing any more. In the old days I always used to have casual work. I couldn't get through it at one time, but now I can't get enough. If there was still a living wage on the land, I should prefer the land, but as things are now there isn't a living wage. My home costs four times as much as it used to. The rents around here are much more than they are in Oxfordshire, or any of the southern counties, although they're getting practically the same wage. If it hadn't been for the pit coming here, there wouldn't have been so much demand for the houses, and rents wouldn't have been so high.[29]

[26] Butterley E4/14 198.

[27] Mr Hedge in *Manor to Mine*, an unpublished collection of oral statements compiled by Robin Whitworth, and dated 16 Nov. 1936. Discovered in the archives of the Butterley Company, now stored in Derbyshire County Record Office.

[28] Ibid., Sam Brown.

[29] Ibid., Spider Parkin.

Mr Thompson, who made monumental tombstones, said:

I wish the pit hadn't come. I've nothing against it in a sense, but we were a very happy community without it. There wasn't nearly so much competition as there is today . . . I myself pay six times as much in rates as I used to before this pit came. That's because they've had to modernise the roads to take more traffic, and put in new sanitary works because there are so many more people. Of course we have got some advantages for our money, but we pay sadly too much in rates.[30]

Such comments received short shrift from those miners contributing. Mr Varley showed himself to be another believer in progress:

We as coal workers can understand the feelings of the villagers, and sympathise with their point of view. People seem to blame us for the general bustle and speeding up of life which is taking place all over the world. We did not come here to spoil the beauty of the Dukeries, we're here because we've got to make a living, and that as far as I know, is the only reason, and the whole of the reason, why any man works in a coal mine. We are sorry, but however sorry we may be, the work of the world has to go on.[31]

This conflict of attitudes was felt by others in both old and new parts of the villages. A resident of Edwinstowe remembered the long-standing divisions there:

Well, they didn't want to mix with us, really. They were they and we were us, and they tended to look down on us. I think that a lot of them thought that the colliery people would bring the tone of the village down . . . it seemed almost as though High Street at Edwinstowe seemed almost a dividing line in the vlllage at one time. The old side of the village was one side, and the new side, you know, the colliery side, was on the other side of High Street, and they just didn't like us in the shops. I can remember during the War going into one of the shops on High Street, and of course we had coupons for sweets, with the rationing, and I can remember going into one of the shops at the top of High Street, it was Woodhead's actually, he owned a toffee shop then, and unless you were old village, he didn't want to serve you. If you asked for a specific thing, you knew somehow that he'd got these—whatever it was you wanted, sweets, chocolate; I can remember Nestle's chocolate was the thing he always held on to—and he didn't want to sell it to the colliery children, you know.[32]

[30] Ibid., Mr Thompson.
[31] Ibid., Mr Varley.
[32] Interview, Barbara Buxton, Forest Town (Edwinstowe).

In some ways the hostility between the 'new villagers' and the 'old villagers' has never died out:

RJW And you were from the colliery village?
BB Yes, I was colliery village. I mean my father went into Edwinstowe in '29, we moved in in '30 when I was still a baby . . . I mean I went to Edwinstowe and I wasn't two. And I left in 1972 and to some of the Old Edwinstowe people I still wasn't an Edwinstowe person. Well, I think when you've spent over forty years in one village, to me you've got to be an Edwinstowe person. I mean I was born in Derbyshire, but all the same I'd lived there, you know, all my life![33]

This view was not unique:

RJW How did people regard each other, you know, the inhabitants of the old village?
JHG Didn't accept us. We're still foreigners to the old villagers, oh yeah.
RJW And what did the miners think of the residents of the old village?
JHG Not a lot. Not a lot.[34]

The same seems to have been true of Ollerton; the former Parish Council Clerk, J. E. Smith, said that 'Well, you mixed, but I think you can understand that in an old village like that, you've got to live there twenty or thirty year before you become a neighbour.'[35] Indeed, as Barbara Buxton discovered, it was possible for people who had lived in the village for the whole of their life to be regarded with suspicion and distrust by the old villagers—a clear example that there were social divisions that time alone could not heal. Like coloured people living in Britain in the twentieth century, the mining families were regarded as immigrants even if they were born in the area.

One reason for the distance and the hostility between old and new villagers may have been that the facilities provided for the miners in the new model villages were not open to the residents of the old villages, and tended to separate the two populations, breeding suspicion and resentment.

It is scarcely surprising that the old villagers failed to make

[33] *Idem.*
[34] Interview, J. H. Guest, Edwinstowe.
[35] Interview, J. E. Smith, Ollerton.

satisfactory social contacts with the miners; in the medium-sized communities like Edwinstowe and Ollerton, there was a well-established 'old village' social life to turn to, but in Bilsthorpe (1921 population 134): 'There was very very little contact. One has to remember what Old Bilsthorpe was like, I mean Old Bilsthorpe was only, what, maybe four or five small farms and the church had got a few cottages, and a few farm cottages . . . so therefore when Bilsthorpe come into being it more or less overcrowded them.'[36] Few societies or organizations crossed the dividing line between the villages. One perhaps predictable exception was the Bilsthorpe Conservative Assocation.

However there were elements in the existing Dukeries society which did welcome the coming of the pits. Most crucially, the aristocratic landowners who had dominated the region up to the 1920s were by no means entirely opposed to the exploitation of the concealed coalfield. The attitude of the Dukeries aristocracy towards the arrival of the coal industry exhibits an interesting dualism. They were reluctant to allow coal mines and coal miners within a couple of miles of their mansions lest they threaten their traditional way of life, the peace and tranquillity of the forest, the hunt, and their political control of local society. Yet on the other hand, squeezed by death duty demands and increasing expenses, they had to recognize that they must 'move with the times' and collect the substantial coal royalties which would accrue from the exploitation of the profitable new field.

The attitude of Earl Manvers of the Thoresby estate is a case in point. At first he was clearly hostile to the idea that his estate's mineral wealth might be tapped. Three quotations from letters written by Manvers's agent R. W. Wordsworth illustrate the landowner's views in 1913:

To J. A. Bell of the Rufford estate:

I have no doubt that you are aware that Lord Manvers is very averse to encouraging a further development of minerals in any way round and

[36] Interview, H. Tuck, Bilsthorpe. Note how a miner refers to the colliery village as 'Bilsthorpe' and the old village as 'Old Bilsthorpe'. This would be regarded as a solecism by an 'old villager', who would always refer to the colliery villages as 'New Ollerton', 'New Clipstone', etc.

about Ollerton, and he has been under the impression that, in this district, at any rate, Lord Savile was of the same mind.[37]

To S. A. Smith of Edenthorpe Hall, Doncaster:

Lord Manvers has no wish at all to develop his minerals on his North Notts. property, and has already declined to entertain doing so to other applicants.[38]

To C. A. Jeffcock of Sheffield:

Lord Manvers is more than ever determined to have nothing to do with developing coal in the Thoresby neighbourhood. It would certainly involve his giving up living there if he did so; and there is nothing which puts him out more than to have such working suggested to him.[39]

Gradually this firm resolve was to be broken down. During the First World War various coal companies continued to show an interest in the Thoresby minerals. Mr Wigram of the Staveley Company met Manvers's agent on 5 September 1916, and 26 March 1917. But after the latter meeting, Manvers withdrew from the negotiations: 'Perhaps, after the War, his Lordship might be willing to reconsider the question, but until that happy event takes place, he has decided not to let the area.'[40] When H. E. Mitton of the Butterley Company wrote in April 1918 offering to work the coal under the Thoresby estate from Butterley's new Ollerton Colliery, he too was sent away with a flea in his ear: 'At present, Lord Manvers has no intention of letting the coalfield.'[41]

By May 1919, however, Manvers's colliery agent C. R. Hewitt was advising that colliery development should take place on the estate:

We have pointed out to his Lordship that it would be possible but not in the best interest of the Estate for the coal to be worked to some of the adjoining Collieries. The splitting up of the Minerals without any sinking on the Estate would not give his Lordship the same security as a fully equipped Colliery on his own land. In our opinion the best development scheme involves a new winning and Colliery Village on the Thoresby Estate, similar to the arrangements already made for Lord Savile at Rufford, Bilsthorpe and Ollerton, where there will eventually be three large collieries in operation.[42]

[37] Nottingham University Library, Manvers of Thoresby Papers Ma 2C 208, 6 Aug. 1913. [38] Ibid., 8 Aug. 1913. [39] Ibid., 21 Oct. 1913.

[40] Manvers Ma 2C 132, Hewitt to Coke, Turner, solicitors of Nottingham, 26 Mar. 1917. [41] Ibid., H. D. Argles to H. E. Mitton, 9 Apr. 1918.

[42] Ibid., Hewitt to Johnson, Raymon Barker, 27 May 1919.

A memorandum drafted in June 1919 was still listing the arguments against allowing coal-mining to come to Thoresby:

(1) There were already four other pits proposed for the area (Bilsthorpe, Ollerton, Clipstone, and Blidworth) and one suspended by war regulations (Harworth, where sinking operations had been undertaken by a German company).
(2) The development of these pits had been deferred because of adverse economic conditions.
(3) Development 'would involve considerable interference with the amenities of the district, because Thoresby Forest is renowned for the beauty of its woodland scenery, and large areas are unenclosed, and the owner wishes them so to remain for the enjoyment of the public. On that account it is likely that Thoresby will remain undeveloped for many years to come.'
(4) Drawing coal from under the Thoresby estate from new shafts sunk outside its boundaries was not to be recommended.
(5) The reservation of the coal seams under Thoresby would be in the national interest, as it would not be wise to exhaust all the seams in one district at once.[43]

Nevertheless, despite this concern for the public and the national interest, the economic advantages of allowing the coal companies onto the estate triumphed. While Hewitt recognized that 'to some extent a new colliery will interfere with the amenities of the Estate for residential purposes', with the other new collieries being sunk in the vicinity there would in any case be a change 'in the aspect of the surroundings', so Lord Manvers must 'participate in the benefits to be derived from the developments of the Coalfield'.[44]

Tenders for the right to mine the coal under the Thoresby estate were invited in February 1920, and the Bolsover Company's offer of 8*d.* per ton for Top Hard coal and 7*d.* per ton for other seams was adjudged the best. Although he had agreed to allow the exploitation of the coal resources, Manvers maintained his attempts to minimize the disruption this might cause to his way of life. An early suggestion was that

[43] Ibid., memo by C. R. Hewitt, 7 June 1919.
[44] Ibid., Hewitt to Johnson, Raymond Barker, 27 May 1919.

the railway line to the colliery should pass in a tunnel beneath Sherwood Forest.[45] In October 1923, when the sites for the colliery villages for Ollerton and Thoresby Collieries were being discussed, the agents of the Savile and Manvers estates, John Baker and Hubert Argles, tried to draw up a pact. Ollerton colliery village should not encroach west of the River Maun or Edwinstowe colliery village south of the railway—either of these developments would bring miners' houses dangerously close to Rufford Abbey and Thoresby Hall, the country seats of Lord Savile and Earl Manvers.[46] If Argles allowed the Edwinstowe village to be built as at first proposed, half a mile from Rufford Abbey's main gate, 'retaliation' might be in order.[47]

A further example of Manvers's continued reluctance fully to accept the industralization of the Dukeries may be seen in his protracted efforts to prevent the new mine being sunk on his land from being named 'Thoresby Colliery': 'Lord Manvers does not wish the new colliery near Cockglode to be known as the "Thoresby Colliery". He would prefer it to be called the "Edwinstowe Colliery".'[48] At least popular parlance could be kept from the Earl's ears: 'I am afraid that the new colliery will undoubtedly often be called "The Thoresby Colliery" but Lord Manvers objects very much to this, so will you kindly instruct your staff to call it "Colliery near Cockglode" or "Edwinstowe Colliery" when writing to this office.'[49] But much as Manvers may have wished not to recognize that the colliery had been sunk on his estate, 'Thoresby Colliery' did become its commonly accepted name.

The main intervention that Earl Manvers made in the process of development of the coalfield was his successful campaign to prevent a railway being driven through the Forest, skirting a famous beauty spot. When Argles first saw the railway company's proposed route, he wrote to Hewitt that 'I am afraid it will seriously interfere with the amenities of the neighbourhood and I hope it will be strongly opposed.'[50]

[45] Manvers Ma 2C 208, Argles to Bingley (Bolsover Company), Mar. 1920.
[46] Manvers Ma 2C 133, Baker to Argles, 30 Oct. 1923.
[47] Ibid., Bingley to Argles, 14 Nov. 1923.
[48] Manvers Ma 2C 208, Argles to Hewitt, 5 May 1924.
[49] Ibid., 1 Jan. 1925.
[50] Ibid., 23 Oct. 1924.

As *The Times* reported, Earl Manvers was indeed to lead the organized opposition:

Our Worksop correspondent states that one of the most picturesque parts of Sherwood Forest, that section possessing romantic associations with Robin Hood and Maid Marian, is threatened with destruction by the proposals of the L.M. and S. Railway Company. The company have just deposited plans for the branch railway to serve the new colliery now being sunk between Edwinstowe and Ollerton, and, according to these, the railway will run through Ollerton Corner and Beech Avenue.

Beech Avenue consists of four parallel lines of well-grown beeches, over half a mile in length, and forms probably the most remarkable and beautiful woodland sight in England. Ollerton Corner is a well-known rendezvous for the hundreds of thousands of visitors who go every year to Sherwood Forest. An immense open space, it is covered with gorse and bracken—the playground of the Dukeries.

The proposal has aroused the strongest possible opposition locally, more particularly as the new colliery will be served by the L. and N.E. Company's new railway. This railway avoids the most beautiful part of the forest. Lord Manvers states, in a communication, that he is very much averse to a disfigurement and the ruination of this part of the Forest, which includes the famous Major Oak, in whose hollow trunk 14 men can stand. He states that in the leasing of the mineral rights, he made what stipulation he could to preserve the scenery as far as possible.—'I feel,' he states, 'that the force of public opinion will oblige the L.M. and S. to withdraw their proposals.'

It is understood that Lord Manvers is organizing a petition against the proposal, which, he says, will involve the crossing of two main roads and necessitate the cutting down of the famous silver birches. He suggests that the L. and N.E. and L.M. and S. companies should have a joint line with a connexion to the colliery; but he is utterly opposed to the branch line. We are informed that Lord Savile is also opposing the scheme in the interests of the public.[51]

However, the aristocrats were not always so successful in restraining the tide of industralization in the Dukeries. When the Bolsover Company first asked to be allowed to erect a pulverized fuel plant at Thoresby Colliery in June 1928, Argles refused permission on Lord Manvers's behalf, on the grounds that smoke would be seen from Thoresby Hall.[52] But when the company pressed its claims, pointing out that the Manvers estate derived considerable revenue from the

[51] *The Times*, 16 Jan. 1925. Manvers's campaign was successful. For the coal companies' view of this dispute, see below, pp. 93–4.

[52] Manvers Ma 2C 134, 2 June 1928.

royalties they paid, it was agreed that Hewitt should inspect a similar pulverized fuel plant at Birmingham. The Thoresby estate thereupon agreed to the Bolsover Company's request, and having failed to secure an undertaking that the plant would be dismantled if it emitted any visible smoke, had to be content with exacting a 'steam plant rent' of £250 per annum. They had been bought off again. The chimney proved to be even more smoky than feared, and the Manvers estate was still sending men to observe the plant's smoke output through binoculars and registering ineffective protests as late as September 1933.[53]

The fact was that the coal royalties had become a financial necessity, and worried as they were about the threats to their long-established way of life, the Earl and his agent had no choice but to allow the coal company to have its own way. As early as 1921 Bolsover were paying a minimum rent of over £1,000 per half-year, before even the ground on which the colliery was to be built had been cleared.[54] On the death of Earl Manvers in July 1926, the mineral value of his estate was estimated at £412,737, compared with £100,000 in 1900, before the new Dukeries field had been opened up.[55] By July 1931 Bolsover had already paid over £60,000 in minimum rents to the Thoresby estate, and had bought 154 acres of land at £100 per acre for the Colliery Village in 1924.[56] The list of half-yearly payments in Table 21 shows how dependent Manvers had become upon Thoresby Colliery. In 1927/8 Thoresby Colliery, still not fully productive, accounted for half of the Manvers estate's mineral revenue of £17,185. 17*s*. 11*d*. It is scarcely surprising that Argles wrote to the Bolsover Company in July 1931: 'I shall be very much obliged if you would let me have a cheque shortly for half year's Minimum Rent, as I have heavy payments to meet very soon in connection with Death Duties.'[57]

By the time the pits had achieved full productive status in the 1930s, coal royalties were bringing vital income to all the Dukeries landowners. In addition to the royalties on coal dug (see Table 22), Butterley paid Lord Savile a ground rent

[53] Manvers Ma 2C 136.
[54] Manvers Ma 2C 133.
[55] Manvers Ma 2C 134, Argles to Hewitt, 4 Sept. 1926.
[56] Manvers Ma 2C 133–6.
[57] Manvers Ma 2C 209, 20 July 1931.

Table 21. *Royalties paid to Earl Manvers for Thoresby Colliery, 1926-37 (£)*[a]

	first half	second half
1926	4,000	
1927		4,500
1928		4,000
1929	4,000	
1930		4,086. 10*s.* 9*d.*
1931		3,830. 16*s.* 6*d.*
1932		3,959. 15*s.*
1933	3,750	3,959. 15*s.*
1934		4,091. 9*s.* 3*d.*
1935		4,091. 9*s.* 3*d.*
1936		8,778. 6*s.* 6*d.*
1937		8,881. 16*s.* 6*d.*

[a] Manvers Ma 3A 2, Estate Accounts.

Table 22. *Royalties paid to Lord Savile for Ollerton Colliery, 1932-38*[a]

	first half	second half
1932	3,427	3,285
1933		3,942
1934	4,092	
1935	5,166	4,117
1936	6,192	6,752
1937		5,914
1938		4,998

[a] Butterley F1/1 202, John Baker 1924-36; Butterley B13 188, John Baker—General.

which amounted to £369. 16*s.* 10*d.* in 1936. It may well be no coincidence that within a few years of the nationalization of coal royalties in 1938 most of the Dukeries aristocrats had sold their estates and left the area. Welbeck became an Army staff training college. Clumber House was demolished. Rufford Abbey fell into ruins, and eventually its grounds were converted into a County Council Country Park. Only Thoresby Hall remained as a private house, now largely converted into flats. What is more, the aristocrats lost little social standing as a result of the arrival of the coal industry. The miners themselves treated the landowners with respect.

The owners of the estates were on occasion seen in the village streets:

> When I first came to Ollerton, I can remember some children springing to attention at the roadside on Wellow Road, when the old school was still going, and a horse and carriage was coming down the road, and they stood there, bowed slightly and took their caps off. And I thought for a moment it must be a hearse passing, but no, it was just either Lord or Lady Savile.[58]

The Marquess of Titchfield remained the Conservative MP for the area throughout this period, and there was no concerted movement against him among the miners; the Labour party did not get off the ground in the new mining villages until after 1943, when he succeeded to the title of Duke of Portland and resigned his seat in the Commons.

In addition to this, the landowners and their agents were consulted by the coal companies on many matters, such as the appointment of curates to the new churches, action concerning roads, railways, and schools serving the new villages, and all matters concerning the leasing of land for colliery purposes. The co-operation between the Butterley Company and John Baker, Lord Savile's agent in Ollerton, may be judged by the presence in their files of the duplicates of letters consulting Baker on almost every decision taken about New Ollerton village.[59] Roads were named after Lord Savile

[58] Interview, A. E.Corke, Ollerton.

[59] John Baker JP, land agent to Lord Savile. Address: Ollerton House, Ollerton. Born 1869, Dawlish, Devon. Ed. Lawn Hill School, Dawlish, and County School, West Buckland. Married 1891, Emily Cutts. JP for the county of Notts., 1927; Chairman of Rufford Parish Council; Chairman of Ollerton and Retford Burial Joint Committee from 1902; People's Churchwarden Ollerton Parish Church 1909-56; licensed Lay Reader; assistant on Rufford Abbey estate from 1884, and agent from 1917. Captain and hon. sec. of Ollerton Cricket Club 1884-1934. Died 1956. (*Who's Who in Nottinghamshire*, Worcester, 1935, p. 10; Notts. CRO, PR 14428.)

in Ollerton and Bilsthorpe, while Thoresby Colliery itself took its name from Earl Manvers's estate on which it was sunk. When coal was struck at Ollerton in the summer of 1925, the very first load of 'Ollerton Brights' was delivered to John Baker.[60] When Montagu Wright took over Baker's residence of forty years, Ollerton Hall, in 1924, Baker moved into another Butterley acquisition, Ollerton House.[61] The agent's position in the old village was unchallenged. Butterley's business with small landowners was conducted through him; and he remained the People's Churchwarden at Ollerton from 1908 till 1956, spanning the 'two worlds' of the Dukeries. Attempts were made to disturb the life of the Dukeries estates as little as possible. Thoresby Colliery was the first mine in Britain to be built without a chimney; and when Manvers's agent H. D. Argles[62] complained about smoke from Ollerton Colliery in 1932, Butterley mining agent H. E. Mitton wrote aggrievedly:

> You, I know realise, that the turning of an agricultural district into an industrial one was bound to alter many things in the life of the old order, and that my Company have and do try in every way possible to consider the amenities of the Country and the social life of the inhabitants.[63]

Indeed, the only occasion on which the aristocrats could be seen opposing some aspect of the industrialization of the Dukeries with any vigour concerned the new railway proposed by the LMS; but this resistance involved an outside railway company, not the coal companies with which such happy relations were established.

It could be said that there were two views of the economic transformation of the Dukeries in the 1920s and 1930s. There was the official view, purveyed by the public utterances of the coal companies, the local newspapers, the aristocratic landowners, and innumerable speakers at the openings of schemes necessitated by the growth of population in the area: this pointed to the material benefits stemming from the

[60] Butterley F1/1 202.

[61] Butterley C10/3 100, 'Ollerton House'.

[62] H. D. Argles, born 1879. Ed. Charterhouse and Balliol College Oxford (4th Modern History and BA 1902). Agent to Earl Manvers. Lieutenant 3rd County of London Yeomanry. Married 1923 Lady Sibell Pierrepoint, daughter of 4th Earl Manvers. Member, Southwell Rural District Council. (*Balliol College Register, 1833–1933* (1934), 242.)

[63] Butterley F1/1 202.

arrival of the coal-mining industry, and characterized this revolution as 'Progress'. This view was based on the interest of the coal-owners in reaping the rich rewards which were expected to flow from this new coalfield wealthy in resources, an expectation later proved to be largely justified, as the pits have remained the most profitable in the United Kingdom, and have subsidized the very existence of many less viable mines since nationalization. The prospect of the coal royalties won the landowners over to approving the economic changes and they had little need to fear a social revolution which would sweep away their prestige and standing in local society, for the coal-owners' 'model' villages were characterized by a respect among the miners for the aristocratic hierarchy, a respect encouraged by the companies themselves, who had every reason to preserve the principle of 'due order and degree'. So the old élites united with the coal companies to bring mining to the Dukeries, and in an age when central planning was not prevalent, no powerful opposition was heard.

In the face of this alliance, only murmurs of discontent could be heard, reflected in one or two newspaper articles and many memories, of new and old villagers alike, memories of distance, suspicion, and hostility, of fear of the newcomer and the unknown. But if the arrival of mining in the new Nottinghamshire field was organized, efficient, and well planned, the reaction against it was disorganized and negative. Unable by long habit of deference and dependence to blame the landowners or their friends in the coal company management for the arrival of the coal industry, the old villagers showed their resentment by hostility to the miners individually and *en masse*. But this kind of opposition could get nowhere; as time passed the new situation was assimilated, if not willingly. As Robin Whitworth of Ollerton concluded in *Manor to Mine, a Programme of Contemporary Contrasts in Village Life*: 'And so the pit wheel turns, and the men of village change. Chaos and order alternate. Men turn the wheels, and wheels mould the men.'

4. Company Villages

'A new industrial feudalism is being erected in the Dukeries . . .'[1]

The sinking of a new coal mine is a major undertaking requiring a considerable investment and an initial outlay of capital that the entrepreneur hopes will be repaid by many years of profitable coal extraction. It is clear that this investment must be protected by attention not only to the technical aspects of raising and selling the coal, but by a concern which extends beyond the pit to the many extraneous factors which may promote or threaten profitability. Where a new coalfield is being opened up, especially one which consists of deep, highly mechanized pits like those of the Dukeries area of Nottinghamshire in the years between the wars,[2] the owners and management have much to gain and much to lose. In 1922 it was estimated that the investment required for the sinking of Ollerton Colliery would be one and a quarter million pounds, without considering the housing scheme for the colliery village.[3] The evidence of H. E. Mitton of the Butterley Company to the Samuel Commission stated that expenditure on the sinking of Ollerton Colliery to 30 September 1925 had reached £489,922. The estimated final cost of sinking and the purchase of coal royalties was £1,046,000 and of housing a further £550,000, making an investment of £1,596,000 in the new colliery, or £1,960,551 after repaying interest on loans.[4] In 1928, E. J. Fox, the Managing Director of the Stanton Company, said that Bilsthorpe Colliery had cost three-quarters of a million pounds and the colliery village a further half million.[5] When the entrepreneurs involved were the highly professional private coal companies of the new Nottinghamshire field, their

[1] *New Statesman*, 24 Dec. 1927. [2] See above, ch. 1.
[3] Butterley E3/4 145, 'A. Leslie Wright'.
[4] *Royal Commission on the Coal Industry* 1925–6, vol. iii, appendix 31, p. 324. [5] *Nottingham Guardian*, 3 Aug. 1928.

concern to ensure that this initial investment was not misplaced could lead to a level of company influence over the lives of the miners and their families usually associated only with the company towns and mining compounds of the USA or Africa.

In the first place, it is well known that coal companies regarded the provision of housing for their employees 'at the pit gates' as an essential element of their sinking costs. P. H. White estimated that companies were prepared to allocate 38–48 per cent of their total capital investment to ensure that a guaranteed supply of labour could be attracted to the pit and then housed far enough away from the temptation of employment at other mines. This is why company villages were often built as near to the pit as possible without regard to the social or environmental hazards or the location of existing communities.[6] The Royal Commission on the Coal Industry of 1925 reported that:

> The special conditions that attach to the mining industry have their effect upon housing. In most other industries the undertakings can, as a rule, place themselves in some town where housing for workers already exists. In mining, if travelling by the workers is to be avoided, the houses must be placed where the undertakings are, and the undertakings must be placed wherever coal is situated.[7]

There was a particular emphasis on the construction of colliery-owned accommodation in Nottinghamshire after the First World War. Fifty-four per cent of the new company housing erected between 1918 and 1925 was to be found in the East Midlands field, although it only accounted for 17 per cent of employment in the British industry.[8]

Several new mining villages were built in Nottinghamshire in the 1920s and 1930s, all consisting almost entirely of company housing. For example, the history of the building of a village of 832 houses for the Butterley Company at Ollerton between 1922 and 1932 can be traced, as the company's business archives have been deposited in Derbyshire County Record Office.[9] The first tenders for the construction of ten

[6] P. H. White, 'Some Aspects of Urban Development by Colliery Companies', *Manchester School*, 23 (1955), 274.

[7] PP 1926 xiv, Royal Commission on the Coal Industry, 213.

[8] White, *Manchester School*, 1955, p. 271.

[9] Derbyshire CRO NCB Inherited Records N5.

semi-detached houses 'which we wish to erect forthwith at Ollerton to accomodate [*sic*] the clerical staff who will be there in the early days of the sinking'[10] were invited on 25 July 1922 and awarded on 30 September of the same year to Messrs. Coleman and Blackburn of East Kirkby, who quoted a sum of £6,900.[11] This proved to be the beginning of a long association between Butterley and Coleman. Building continued at a pace which caused great problems for the local government authorities responsible for educational, transport, and sewerage facilities during the period of the growth of the new village. In 1929, for example, a phase of building 100 new houses at Whinney Lane, New Ollerton, was first mooted on 4 March, the contract let to Coleman in mid-April and all were occupied by 9 September.[12] The cost was £49,140, towards which the Southwell Rural District Council contributed £50 per house under Chamberlain's housing subsidy scheme. Indeed, the summer's work might have been completed even more quickly had it not been for the red tape involved in securing the subsidy. The company's attitude was clear. The mining agent, Eustace Mitton, visited the building site on 10 June: 'It was a beautiful evening at 7.30, and there were only two men smoking cigarettes on the Site, and it did occur to me that the Contractor is wasting an enormous amount of valuable time.'[13]

By 12 January 1932, 1,301 of the men employed at Ollerton Colliery were living in company houses in the colliery village, and only 453 living elsewhere. A total population of 3,991 inhabited the 832 houses of the colliery village.[14] Even then Colliery Manager Montagu Wright was requesting that 100 more houses should be built, as forty men had their names down for houses, fifty more miners were required at the pit, and the company could not attract and keep 'the right kind of men' unless accommodation could be provided for them in the village.[15] In fact no more workmen's houses were built at Ollerton until 1941, when a licence was issued by the Mines Department for fifty more, which completed the Butterley Company's building programme.[16] After the

[10] Butterley H3/6 225, H. E. Mitton to John Baker, 6 July 1922.
[11] Butterley H3/6 225. [12] Butterley H3/3 224. [13] Ibid.
[14] Butterley H3 224. [15] Ibid., 5 Jan. 1932. [16] Ibid., 17 Jan. 1941.

war, Ollerton was to grow further with the development of a second pit, at Bevercotes, by the NCB.

At Harworth too, building proceeded at a rapid pace as the colliery village was constructed in the 1920s. The first street to be constructed was Colliery Row, for important workmen who had to live on the site of the colliery because they were responsible for twenty-four-hour maintenance. There followed Scrooby Road, mainly for officials, and six 'blocks' of miners' dwellings.[17]

The inactivity of local authorities in the new colliery villages in the inter-war period forms a striking contrast with the building programmes of the private coal companies. At Ollerton three pairs of 'undistinguished'[18] council houses were built in 1940 at a total cost of £2553. 17*s.* 2*d.* and two pairs were constructed at Bilsthorpe at the same time. Such was the full scope of the involvement of local government in the growth of the colliery communities in the Dukeries before nationalization.

It hardly needs to be stressed that the near monopoly of housing in New Ollerton possessed by the Butterley Company offered them a considerable amount of control over their workmen. As Charles Graves put it in his article in *The Sphere* in 1932: 'The company loses £1,500 a year over its housing scheme, but the tranquillity and content of the employees make it a cheap investment.'[19] It was made quite clear in the rental agreements of the houses that employment at the pit was a condition of occupation, and that the lease was weekly.[20] Dismissal meant almost instant eviction. Since no one dared take as a lodger a 'trouble-maker' sacked by the company, this meant that effectively Butterley could expel anyone from the village. The Butterley Company management frequently undertook their own censuses of the colliery village to make sure that only those 'useful to the Company' were resident in company houses.[21] There were also a number of other disciplinary conditions which the company imposed on tenants, as will be seen below.[22]

[17] Interview, Charles Stringer, Harworth.
[18] PRO HLG 49/1297.
[19] Charles Graves, 'A Miner's Life Above Ground', *The Sphere*, 23 Apr. 1932.
[20] Butterley H3 224, leasing conditions, Jan. 1940.
[21] Butterley H3 224.
[22] See below, ch. 4.

This was true in all the colliery villages of the new Nottinghamshire field, where the companies could start from scratch and provide all the housing for the community. At Edwinstowe, the Bolsover Company had built 497 houses for workers at Thoresby Colliery by 1931.[23] The Industrial Housing Association built 821 houses for the Newstead Company at Blidworth Colliery village between 1924 and 1927.[24] The Stanton Company erected 470 houses for its new pit at Bilsthorpe between 1926 and 1929;[25] 648 houses were erected for Bolsover at New Clipstone, and 1,100 for Barber-Walker at Harworth in the same period.[26] All of these developments were seen as model villages, although it will be argued that they owed little to the ideals of the garden city movement of Ebenezer Howard.[27]

They were planned communities, of symmetrical layout with semi-circular closes and broad avenues.[28] (The nature of the layout can clearly be seen in the plan of Edwinstowe colliery village (Fig. 3). The plans of the IHA's Blidworth village (Fig. 5, chapter 12 below) should also be consulted.) The houses were in the main not terraced but semi-detached; they enjoyed bathrooms and a hot-water supply. At Ollerton the Butterley Company was very proud of its system whereby water was heated at the colliery and circulated round the village in external pipes. There can be no doubt that the new housing was superficially more attractive than that in the old coalfields from which the immigrant miners had come, yet the prevalence of company housing undoubtedly did contribute towards the feeling that these new villages were repressive company towns in which a false step might mean eviction and departure from the village. Many of the houses in the new communities were of standard and monotonous design. N. Summers judged that 'the physical standard of

[23] Thoresby Colliery Manager's Annual Report, 1931.

[24] Sir John Tudor Walters, *The Building of Twelve Thousand Houses* (London, 1927), pp. xv–xxiv.

[25] C. M. Law, 'Aspects of the Economic and Social Geography of the Mansfield Area', Nottingham University M.Sc. 1961, vol. 2, p. 163.

[26] G. C. H. Whitelock, *250 Years in Coal* (Derby, 1955), p. 72.

[27] See below, pp. 261–2.

[28] In the case of Ollerton, each road had to be at least 36 ft. wide—Butterley H3/1 224.

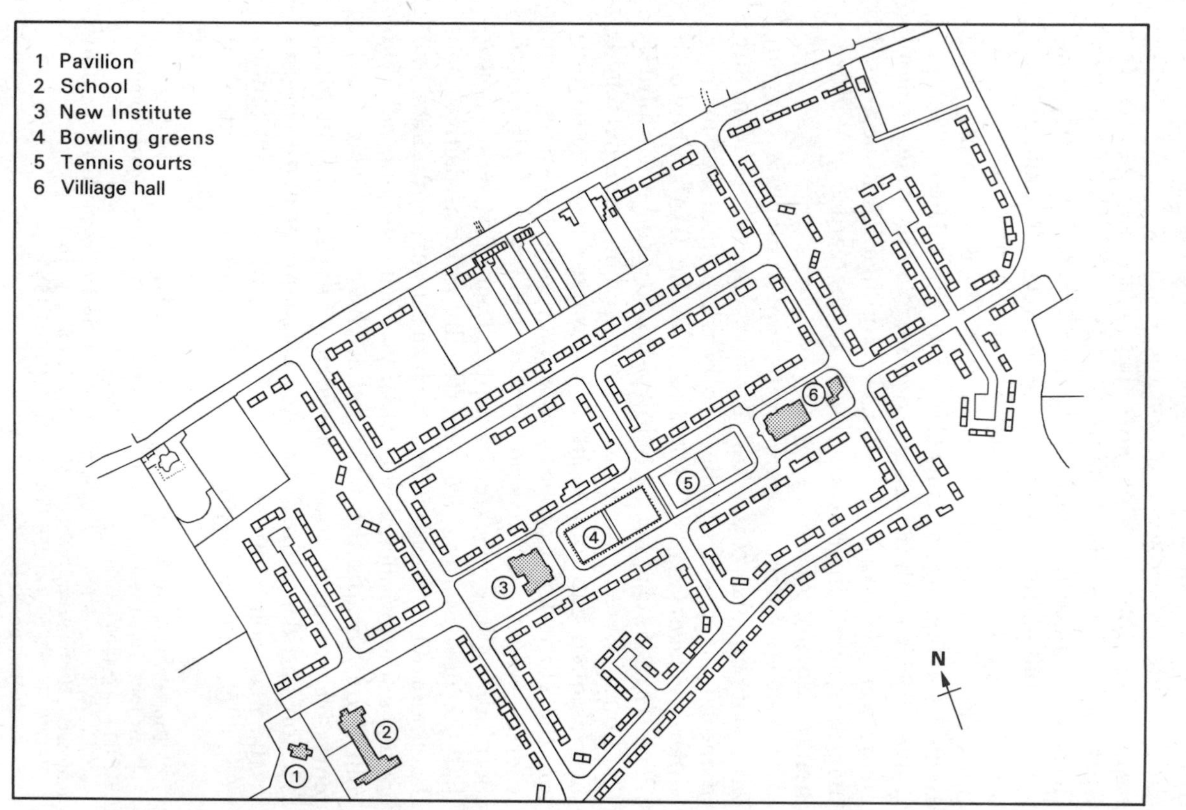

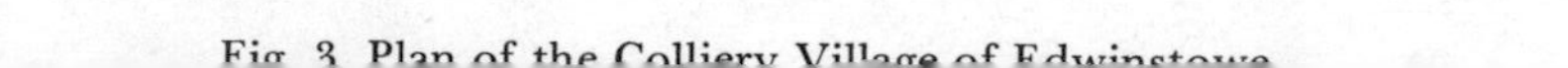
Fig. 3. Plan of the Colliery Village of Edwinstowe

housing is high for the mining area before the last war although the neo-georgian manner here lacks conviction and scholarship' and he condemned 'a dull monotony of lay-out'.[29] In the case of Blidworth, 'the whole shopping environment is the makeshift planning of a mushroom growth which has never been matured', and contained unimaginative architecture such as that of the company-built pub, the Forest Folk, 'a grimly ponderous building of the '30s'.[30] Nevertheless, for all the lack of variation in house design, important social distinctions were recognized by the companies as they constructed their villages.

At Ollerton forty-seven sinkers' bungalows were built in a kind of shanty town for the predominantly Irish labourers employed in the 1920s before the pit reached productive status (see Fig. 4). In fact they were used for many years and the last one was not demolished until 1978. It is clear that the residents of the prefabricated bungalows were looked down upon by the other inhabitants of the village long after the sinkers had moved on:

> I can remember kids at school, if you knew they came from the bungalows, you tended to look down on them. It was all the sort of scruffy ones who were associated with the bungalows . . . well, until I was sixteen, I always equated bungalows with scruffiness, I mean I think bungalows are marvellous now, but if somebody said he came from a bungalow, oh dear![31]

Such sinkers' dwellings were also found at Harworth and at Bilsthorpe, where 'temporary' dwellings were erected by the Stanton Company at a cost of £26,499 in 1924/7. When the Southwell Rural District Council sanctioned their construction, they attempted to add the proviso that Stanton should give an undertaking to take the bungalows down at the end of ten years. However, the company refused, pointing out that the Council had a statutory right to require Stanton to take the houses down at any time if it was proved that they were unfit for occupation.[32] The Council then agreed that they would not ask Stanton to take the buildings down

[29] N. Summers, 'Problems of Visual Environment in Nottinghamshire', Nottingham University Ph.D. 1966, vol. 2, p. 14.

[30] Ibid., vol. 2, p. 13.

[31] Interview, Cyril Buxton, Forest Town (Ollerton).

[32] Stanton Company Colliery Committee Minute 23, 16 June 1924.

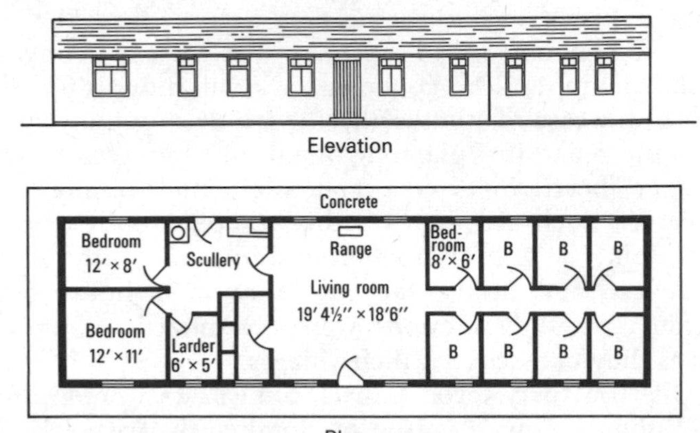

SINGLE QUARTERS

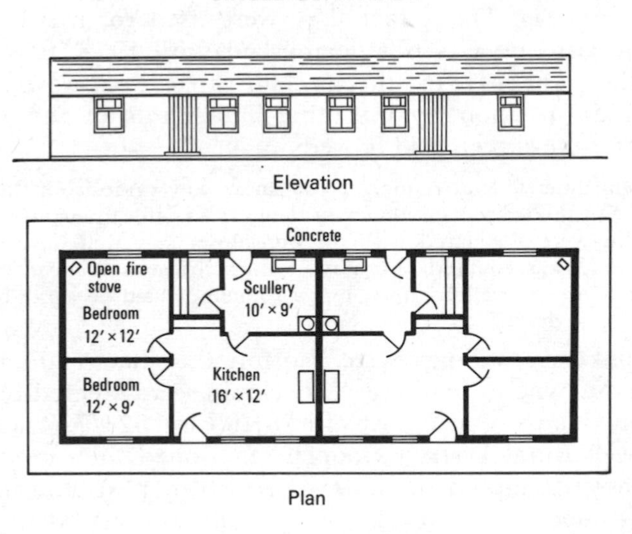

MARRIED QUARTERS

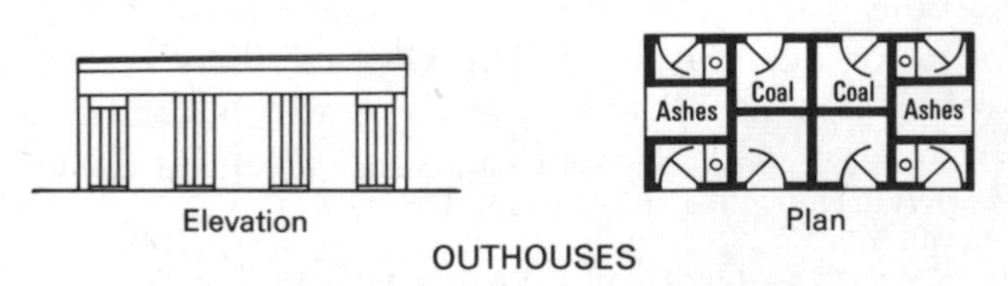

OUTHOUSES

Fig. 4 Sinkers' Huts, Ollerton Colliery

before 1 January 1935 at the earliest,[33] and in fact the shanty town remained complete until the 1960s.[34]

At Edwinstowe, no bungalows were built, but the sinkers were concentrated in a street destined for a poor reputation:

RJW Who were the sinkers? Were the sinkers an entirely different type of person?
DT They only dug pit.
FT They were all housed on one avenue.
DT On Fifth Avenue.
GC Where the Geordies were![35]
RJW Where did the sinkers come from?
GC All over. They were Irish, they went around the country sinking.
FT Irish, Rechabites, Jacobites, all sorts. They did nothing else.
RJW I know at Ollerton they were mainly Irish sinkers, they put them in a kind of shanty town . . .
GC Well, there was no shanty town here as such, but 5th Avenue was—
FT A disaster area![36]

On the other hand, semi-detached houses were provided for the officials of the colliery, which were larger than the standard types provided for the workmen. On 5 July 1922, A. Leslie Wright, Managing Director of the Butterley Company, wrote: 'I suggest that we should erect half a dozen good semi-detached villas for these Officials, commencing at once . . . I could not expect these Officials to reside in temporary Hutments, such as I intend putting down for the sinkers.'[37] At Edwinstowe all the officials had to live in one avenue, First Avenue. When one interviewee was promoted to overman he had to move house within the colliery village from Third to First Avenue.[38] At Bilsthorpe, six pairs of officials' houses were constructed at a cost of £1,984 per pair.[39]

Still further up the social scale, the Butterley Company built substantial detached residences at Ollerton for the curate-in-charge and the doctor.[40] These were houses suitable for the professional classes. Each cost over £1,700 to

[33] Ibid., Minute 30, 8 Sept. 1924.
[34] Interview, M. Morley, Bilsthorpe. [35] See above, p. 47.
[36] Interview, Frank Tyndall, Dorothy Tyndall, George Cocker, Edwinstowe.
[37] Butterley E3/4 145. [38] Interview, J. H. Guest, Edwinstowe.
[39] Stanton Company Colliery Committee Minute 37, 13 Oct. 1924.
[40] Butterley H3 222.

build,[41] and the doctor's house was provided with its own surgery. The land for these dwellings was given free by Lord Savile. The doctor appointed, Dr Cuddigan (a Roman Catholic, like Monty Wright) often proved a valuable ally for the company in compensation cases, on some occasions declaring men fit for work when the doctor nominated by the union had deemed the men concerned liable for payment by the company.[42] The Bolsover Company also treated its doctor in Edwinstowe well:

FT The Bolsover Company had one doctor, Dr Lloyd. Nigel Fitzroy Lloyd, that was his name.
DT On Sandy Lane.
GC I knew him well, he ran an Armstrong Siddeley, which was a company car.[43]

A doctor's house was also built at Bilsthorpe, in 1933 at a cost of £1,100 to the Stanton Company, with the condition that the doctor should maintain his practice for a given period.[44] The company tried to tie the doctor to supporting their case in any compensation case that might arise at Bilsthorpe Colliery. The Stanton Company Colliery Committee heard on 11 October 1926 that:

The Secretary reported that an offer of £100 had been made to Dr. Robinson as a retaining fee for the year ending 30th September 1927, with the stipulation that he would not act against the Company in any Workmen's Compensation Cases, and would perform such medical services at Bilsthorpe as the Company may require. Dr. Robinson replied that he would be glad to accept the Company's offer if the condition with reference to Workmen's Compensation cases could be omitted, and the services required by the Company would not include any which would prejudice the goodwill of his patients.[45]

At the top of the local hierarchy, a dwelling had to be provided for the colliery manager in the new pit villages. At Ollerton this problem was solved in 1925 by the renting from Lord Savile of Ollerton Hall, a mansion previously occupied for forty years by the local aristocrat's agent, John Baker.[46]

[41] The curate's house cost £1,725. 17*s.* 7*d.*, the doctor's house £1,800—Butterley A2/2 14.
[42] Butterley C7/7 'Compensation'.
[43] Interview, Frank Tyndall, Dorothy Tyndall, George Cocker, Edwinstowe.
[44] Stanton Company Colliery Committee Minute 928, 10 July 1933.
[45] Ibid., Minute 252.
[46] Butterley A2/2 14.

The colliery manager, Montagu Wright, a member of the family which owned the Butterley Company, moved into the Hall. It was an ideal spot from which he could keep an eye on all the village activities, and, perhaps, inherit some of the deference traditionally due to the lord of the manor and his representative. Montagu Wright later became Managing Director of the Butterley Company, but he lived in Ollerton Hall until 1947, and the nationalization of the coal mines.

At Edwinstowe too, management took over the most substantial houses in the village: Bolsover Company mining agent T. E. B. Young (later Sir Eric Young) occupied Edwinstowe House. In 1923 Bolsover acquired Edwinstowe Hall.[47] However, here a new house was built for the manager of the colliery:

FT Woodward, Charles Edward Woodward, he was the first manager.
RJW And he lived in the village, did he?
FT Yes, he *was* the village, in that sense!
RJ Whereabouts did he live?
GC The manager's house, it was built for him down Ollerton Road.
FT Down Ollerton Road, that's a few yards from the crossroads, about as far as you've come this way, the other way.
RJW Towards the pit?
FT Yes. He could see the pit. He could see the wheels going round, and if they weren't, he wanted to know why![48]

At Bilsthorpe, not only was a new manager's house built, but the hierarchic layout of the colliery village was particularly well developed:

RJW And where did the managers live, did they live in the village?
HT Oh yes, in special houses.
RJW Do any of these houses still stand?
HT Oh yes, yes, still stand.
RJW Whereabouts are they?
HT Well, where the colliery is now, our manager, who is the present manager, Mr Watt, he lives in the original house that was built. If you went back towards the pit now, it's a big house, stands all on its own, on a rise of ground, as you go towards Ollerton.
RJW And it was built as a manager's house?
HT Oh yes, oh yes. And the other officials' houses, the undermanager and all the lot, they were built near the colliery, if you went from

[47] Bolsover Company Jubilee Souvenir 1889–1939.
[48] Interview, Frank Tyndall, George Cocker, Edwinstowe.

here to the colliery, the last houses, row of houses, which was about eight, maybe ten, they were specially for management, and you had your head electrician, head mechanical engineer, and your undermanagers, etc., and they used to live in them houses.

RJW And in that time you had a kind of hierarchy?

HT Oh yes! There were different scales, yes. Even in the village itself for miners there were.

RJW How were they–

HT Well, if you were a chargeman, if you were what they call, the term in those days were 'butties', used to call them butties, chargemen. And if you were well in, you got one of those houses. . . . Start from the top of the hill there. Them colliery houses, nearly every street is different. Some have got parlours, as they call them, dining rooms if you like, we used to call them parlours. There were some with bathrooms and some without, and all sorts of different things. Of course when you got a job in the village you had what house were vacant, and that was it, full stop. But if there was a nice house in the street, well very often the people what worked here went in and said 'Well, I'd like that house, gaffer', and if you were well in, good chap, good works record, you used to change. And so the bloke that came in from Durham had to have the one that weren't quite so good. Of course they were all modern houses you know, that were all newly built.

RJW And which streets were the best houses on, and which streets the worst houses?

HT Well it started from Eakring Road, which was deputies' row, and then worked back towards where we are now, it worked back. You'd got Eakring Road, which was deputies' row, if you like, officials. Then you'd got Savile Road, which was for the good chargemen if you like and all such as that, and then you'd got Crompton Road, Cross Street, Scarborough Road, more or less for, well, anybody.

RJW So as you got further away from Eakring Road, it gets sort of lower down the scale?

HT Oh yes, yes.[49]

The same was true at Harworth. When asked whether the colliery officials were given better houses, one respondent said:

Oh yes, aye, they were like NCOs, they were given better quarters than what the colliery workers were. If you take notice when you go through the village and you go down Scrooby Road, you'll see straight away from Shrewsbury Road down to Waterslack Road, the houses are much larger and better built. Those are the officials' houses, the cashier of the colliery, the overmen of the colliery, and the deputies. In fact there was

[49] Interview, Herbert Tuck, Bilsthorpe.

one house down there that the company provided for the doctor, which was known as the pit doctor. They even helped a doctor to come to the village and provided him with a house, provided him with a surgery within the colliery yard that's still used as the surgery today, although it owes no allegiance to the colliery whatever.[50]

The social distinctions in the housing in the new villages were reflected in the rateable values of the coal companies' houses. At Bilsthorpe, the thirty-six deputies' houses on Eakring Road and Church Street were rated at £13 in 1939/40; 364 ordinary miners' houses were rated at £12, and six houses for senior colliery officials on Eakring Road at £20. At Ollerton, 49 bungalows were rated at between £5 and £12; 570 houses at £12 and 44 officials' houses on Forest Road at £17.[51] Montagu Wright's Ollerton Hall was assessed at £122 in 1940.[52] At Harworth the colliery manager lived in a substantial house at the top of a hill which is still known as Wright's Hill, having been named after the first manager of Harworth Colliery, William Wright:[53] 'Mr Wright lived there, and we used to say "we're going down Wright's Hill" because he were gaffer and he were Him at that time of the day, he was the king here.'[54]

When it came to the naming of the new roads, a variety of ideas occurred to the coal companies. At Ollerton the broad avenues were named after types of tree, such as Oak Avenue, Pine Avenue, and Sycamore Road. An exception was made for the superior semi-detached houses reserved for officials: on 9 June 1925 the Butterley Company board decided that official's houses should be known as Savile's Row in honour

[50] Interview, Stan Morris, Harworth.

[51] Southwell RDC rate books, 1939-40, Notts. CRO DC/SW/3/1/1.

[52] Butterley A2/2 14.

[53] Cyril Murphy, Manager of Harworth Colliery, *Harworth Colliery 1924-74* (typescript, unpublished), p. 3. William Wright was born in Lancashire, where he received his early training. Appointed undermanager of Barber-Walker's Watnall Colliery in 1914, he qualified as a manager in 1915 and became manager of High Park and Watnall Collieries in 1916. He was appointed manager of Harworth Colliery in 1923, two months before coal was reached. He became agent at Harworth in 1943, and continued in that position after nationalization until he retired in 1948 after 58 years' service in the industry. He was a Justice of the Peace, serving on the bench at Worksop; Chairman of Harworth Parish Council; and the Worksop Rural District and Notts. County Council representative for Harworth. (Whitelock, *250 Years*, p. 239.)

[54] Interview, Tommy Jenkins, Harworth.

of Lord Savile of Rufford Abbey, on whose estate Ollerton was built.[55] At Edwinstowe the roads were originally intended to be named after the Bolsover Company's other pits (Creswell, Rufford, etc.) although this never caught on and the unimaginative 'First to Seventh Avenues' became established.[56] At Bilsthorpe, Crompton Road and Scarborough Road were named after Stanton Company directors, and Savile Road again honoured the local landowner. At Langold,

EB You'd Mellish Road, that was your local squire, Mellishes of Hodsock Priory; White Avenue, that's Archibald White of Wallingwells. You'd Riddell Avenue, that's Squire Riddell of Oldcotes; you've Ramsden Road, that's Squire Ramsden of Carlton; Church Street's self-explanatory; School Road, that's self-explanatory; Wembley Road, I always thought F. K.[57] named that, he went to Wembley so often.

JJ No, Wembley started at that time, '23, first Cup Final.[58]

However, the coal companies' property in the new villages was not limited to the housing. In general the companies leased the whole area on which the villages were to be built from the Dukeries aristocrat on whose estate it was situated —in the case of Ollerton, from Lord Savile of Rufford Abbey, who owned 5,000 of the 7,000 acres leased by the Butterley Company from 1922 for sixty years.[59] This meant, of course, that with the consent of the landowner, the company could decide how to dispose of all the land in the colliery villages. Their building went far beyond the mere provision of the houses needed for their workmen and staff. At Bilsthorpe the company owned the local public house, aptly named the Stanton Arms,[60] licensed in 1925,[61] and sold to Samuel Smith's Old Brewery of Tadcaster in 1946.

The Stanton Company also owned and ran the only shop in the new colliery village. When in June 1925 the Southwell Co-operative Society applied to open a shop at Bilsthorpe,

[55] Butterley F1/1 202, 'John Baker 1922-36'.

[56] Interview, Dorothy Tyndall, Edwinstowe; this can also be confirmed from the naming of the roads in the electoral registers, which changed to numbers in 1930.

[57] Frederick Knaton Godber, Cashier of Firbeck Colliery.

[58] Interview, Ernest Baker, Jim Jackson, Langold.

[59] Butterley E3/4 150, '32 Houses Ollerton Public Works Loans Board'.

[60] Notts. CRO DC/SW/3/1/1, Southwell RDC rate books.

[61] Stanton Company Collieries Committee Minute 70, 9 Feb. 1925.

the Stanton Colliery Committee decided that no land should at present be leased for the purpose of erecting a shop but asked the Stanton Board 'to consider a proposal that the Stanton Company should erect and control a shop or store at Bilsthorpe for the sale primarily of food.'[62] In October 1926 it was again deemed inadvisable to offer the Co-op a tenancy,[63] and by February 1927 a company shop had been established which was taking £40 to £50 a week.[64] The manager of the shop also assumed the duties of sub-postmaster from 1 September 1927,[65] and in November of that year the Colliery Committee approved the spending of £300 to enable the store to sell patent medicines to preclude the need for an independent chemist's shop in the village.[66] By June 1928 it was found necessary to employ an assistant to the shop manager and a youth to deliver goods in the growing village.[67] The shop remained the only retail outlet in the village of Bilsthorpe, with the exception of travelling salesmen,[68] until 1935, when the business finally was sold to the Mansfield and Sutton Co-operative Society in the hope that the Co-op would continue to purchase 70,000 tons of coal a year from Stanton.[69]

The company shop could be used as an instrument to control the employees who had to patronize it:

HT And even in them days if you didn't pay your grocery bill, and left it one week, it were stopped out of your wages, because they'd got control, and you couldn't do nowt about it, because if you argued about it you were up the pit lane, got sack!
RJW So they actually stopped your grocery bill at the pit?
HT If you got in arrears, yes.[70]

[62] Stanton Company Collieries Committee Minute 104, 8 June 1925.
[63] Ibid., Minute 252, 11 Oct. 1927.
[64] Ibid., Minute 290, 14 Feb. 1927.
[65] Ibid., Minute 338, 8 Aug. 1927.
[66] Ibid., Minute 363, 14 Nov. 1927.
[67] Ibid., Minute 437, 18 June 1928.
[68] Ibid., Minute 438, 18 June 1928. The Barber-Walker Company were determined to keep even travelling salesmen out of the village of Harworth, and erected a large 'No Hawkers' sign at the entrance to the village (interview, C. E. Stringer, Harworth).
[69] Stanton Company Managing Committee Minute 1528(a), 15 July 1935, in which the shop is described as a quid pro quo for the extension of the contract to supply coal to the Co-op.
[70] Interview, H. Tuck, Bilsthorpe.

Another resident of Bilsthorpe described the effect of the company's ownership of the pub and the stores: 'So what happened, you were working for the company the whole 24 hours. You worked at the pit; if you were sleeping you paid company rent, if you went out for a drink it were their money, if you bought your groceries it were their money.'[71] Another example of the company store, so evocative of American experience, was to be found at Edwinstowe. Inevitably a certain amount of pressure to use the shop was perceived by the inhabitants of the colliery village:

RJW Did they like the miners to shop there, or didn't they mind?
FT Well, it's just one of those things again! If you didn't shop there, where did you shop? 'Cos if you were seen going down the street to another shop, there'd be somebody there telling the deputy, the next morning, 'ah, so and so goes shopping down there, why can't we?', and that got to the manager, balloon went up![72]
RJW Was there any pressure from the company or any suggestion that the people from the colliery village should go to that shop rather than go into the old village?
BB Yes, I don't think it was ever said as such, but the inference was there all the time.[73]

Other company stores were built at Harworth and Langold.

At Ollerton, Butterley decided not to build a store of their own. The Board did suggest setting up a Central Stores selling 'everything the employees require, from a Mouse Trap to a Motor Car, a sort of Selfridges in a glorified state'[74] but H. E. Mitton, the mining agent, felt that this would lead to the miners living on credit, seeing the company as bound to supply their needs without payment. Instead the company leased plots of land for lock-up and permanent shops, which prospective traders had to build for themselves. This explains why the only irregular building development at Ollerton, as at Blidworth and Clipstone, is on the main shopping street.

Butterley did, however, make up for not having a company store in Ollerton by constructing two other remarkable buildings. First, it was decided by the Board of Directors in 1926 that Butterley should build a church in Ollerton 'as a cathedral

[71] Interview, W. Sperry, Ollerton (Bilsthorpe).
[72] Interview, F. Tyndall, Edwinstowe.
[73] Interview, Barbara Buxton, Forest Town (Edwinstowe).
[74] Butterley E4/14 198.

for the new coalfield'.[75] Although this church was not designed by Sir Giles Gilbert Scott, the architect of Liverpool Cathedral, as Butterley had hoped at first, £5,000 was donated towards the construction of St. Paulinus Church in New Ollerton colliery village. As will be seen in chapter 8 below, all the other colliery companies which constructed new villages took a great interest in encouraging religious observance amongst their employees, helping to build Church of England and Roman Catholic churches along with Nonconformist chapels, and taking care to influence the appointment of the curates who were able to work in the colliery villages. Although the companies insisted that they were concerned solely with the spiritual guidance of the miners, their concern with religion fits in with their policy of making every aspect of life in the colliery villages a matter for their scrutiny.

Butterley's second major building investment in Ollerton colliery village, besides St. Paulinus Church, was also apparently far removed from the affairs of the pit itself. It had quickly become clear that miners were reluctant to live in New Ollerton because of the complete lack of employment for female labour—usually this concerned their unmarried daughters, although wives occasionally worked. The village was set in a hitherto rural area ten miles from the nearest town, Mansfield—too long a journey to make twice a day. As early as 19 November 1931 Butterley were making plans to ensure that men were not lost to them because of the lack of this essential supplement to the family income, by attracting them to Ollerton with the offer of employment for the whole family.[76] A plan whereby S. Shephard of East Bridgford would agree to take a factory in Ollerton to provide employment for the girls in the colliery village was floated in 1933.[77] The Butterley Company undertook to protect his labour supply by not selling land within a five-mile radius of the factory. In return, Shephard was to guarantee that at least 65 per cent of his labour force had to be members of the families of miners working at Ollerton Colliery.

[75] This was how the *Southwell Diocesan Magazine* described the Butterley Company's original intentions. *Southwell Diocesan Magazine*, Sept. 1932, p. 149.

[76] Butterley H3 223, Wright to Mitton.

[77] Butterley L1 226.

This scheme fell through, but Butterley did not give up, and in 1936 the coal company contacted Messrs. Hall and Earl, hosiery manufacturers of Leicester, through an intermediary, Lt.-Col. Dawes of the Federation of British Industry in Nottingham. Montagu Wright wrote to Dawes on 26 August of that year: 'I feel certain that once a factory was put up, that the firm who put it up would not regret the step which they had taken, and I feel certain that the Butterley Company would do all they could do to help any firm who proposed erecting a factory in Ollerton.'[78] Indeed so; unable to persuade any firm to build factory in Ollerton, Butterley built one itself. A Mr Palfreyman of Hall and Earl had visited the village in September 1936, and was impressed: 'He liked the cleanliness of it, which, he thought, showed that they must be quite a decent sort of person in Ollerton.'[79] The hosiery factory was built by the Butterley Coal and Iron Company during 1937 at a total cost of £27,155. 13*s*. 8*d*.[80] Hall and Earl rented it at £1,629. 6*s*. per annum. When asked by the Southwell Rural District Council to pay for part of the cost of installation of a sewerage scheme for the factor, Montagu Wright protested strongly: 'The Butterley Co. had no particular interest in factories, and their only reason for getting the factory built, was to help the district by finding employment for the girls in the district.'[81] That this was a little disingenuous is revealed by a letter from Butterley's mining agent to the company solicitor:

> The main reason for the Butterley Company considering the building of a factory at Ollerton is that we are very short of coal face workers, and it is almost impossible to get colliers to come to Ollerton, who have girls working at the place at which they are employed. There is no scope for female labour in Ollerton and district, and we have had quite a few men leave simply because there was no work for their girls.'[82]

By January 1939 Hall and Earl were employing about 250 girls from the new coalfield. Wages were lower than at similar factories in the towns, but then so were travelling expenses. Wright was even putting out feelers about the possibility of another factory being built in Ollerton. He wrote to R. B.

[78] Ibid. [79] Ibid., Mitton to M. F. M. Wright. [80] Ibid.
[81] Butterley L1 226, M. F. M. Wright to the Clerk to Southwell RDC, June 1937. [82] Butterley L1/4 228, J. Bircumshaw to Hull, 30 Nov. 1936.

Edwards of Drewry and Wright, hosiery manufacturers of Nottingham: 'We do not particularly want work to be found for male labour, as the colliery provides the necessary work for them, but we do feel that with the surrounding Villages there is ample female labour available to staff another Factory.'[83]

The Ollerton hosiery factory is probably the best example of the Butterley Coal Company's imagination and vigour in seizing opportunities to protect their investment in Ollerton pit. But their interest and activity in the village were almost boundless. A £5,000 grant was obtained from the Mines Welfare Committee for the building of a permanent Miners Institute in Ollerton, opened in 1928.[84] The company was also concerned with matters not directly within their jurisdiction. They were constantly pressing the inefficient local authority, the Southwell Rural District Council, to expedite the provision of essential road and sewerage services for the village which was growing as a result of colliery development.[85] The need to protect their interests in matters discussed by the local authority led them to make sure that at least one of the representatives of the colliery villages on the Rural District Council was a senior official of the colliery.

Another necessary service was the mineral railway line which would form an outlet for the coal produced at the new pits. Here Butterley combined with other coal companies, Stanton and Wigan, to put pressure on the LMS and LNER, each of which applied for parliamentary Acts to build railways to serve the new coalfield. By threatening to oppose both plans in Parliament, the coal companies managed to persuade the two railway companies to agree on a joint line.[86] In 1924,

> now that new pits are being developed at Blidworth, Bilsthorpe, Rufford, Edwinstowe, Ollerton and Boughton (not to mention other proposed sites), such a railway has become a paying proposition with the result that both the great railway companies, no doubt having got wind of each other's plans, are submitting to Parliament largely identical proposals.[87]

But on 3 April 1925, after much high-level negotiation on the

[83] Butterley L1/4 228, 5 Jan. 1939.
[84] Butterley W2 440.
[85] See below, pp. 145–51.
[86] Butterley P1/2 271.
[87] *Nottingham Journal*, 28 Nov. 1924.

part of the coal companies, both rail giants withdrew their rival bills from the Lords, and in December 1925 a joint bill entered Parliament which received the royal assent in July 1926. Now the coal companies had to move the LMS and LNER in a different direction: by 9 November H. E. Mitton of Butterley was threatening that if no progress was made to start the new line soon, the coal company would introduce its own private bill to start construction itself. After all, as Mitton pointed out to Sir Ralph Wedgwood, Chief General Manager of the LNER:

> The development of the Colliery and the building of the colliery village has been proceeded with energetically and with great expenditure of capital in the belief that the Railway Companies would be equally energetic in providing the promised traffic facilities. You must forgive me for saying that the practical waste of fifteen months above referred to is not what we were reasonably justified in expecting from your and Mr. Morley's companies.[88]

Butterley was equally concerned about the Notts. County Council's provision of schools for Ollerton colliery village.[89]

The concern of the coal companies to ensure that inadequate amenities would not prevent them from attracting the labour force they needed in the newly exploited coalfield is understandable. They paid just as much attention to the welfare and conduct of the workmen when they did arrive. Butterley, like all the coal companies active in the new coalfield, was convinced of the virtue of paternalism. In many ways this took the form of positively encouraging community spirit and of rewarding loyalty to the company. One of the clearest examples of this is the case of sport.

Most of the new collieries possessed strong cricket and soccer teams, and great attention was paid by the management to ensuring the success of representative teams in any competition in which they took part.[90] Bowls players, cricketers, and footballers were put on at the pits as a result of their prowess. Sporting ability was even a criterion adopted by the management when interviewing clergymen for curacies in the new villages. The local newspapers were dominated by

[88] Butterley P1/2 271. [89] See below, pp. 172-3.

[90] See below, ch. 9, for a more detailed treatment of the matters outlined in this paragraph.

sporting stories. Perhaps, as in modern Communist states and Japanese businesses,[91] sport was encouraged by the coal companies of the Dukeries field as a way of inculcating harmony in the pit and in the community, uniting men and management, suppressing 'them and us' conflict theories; as a substitutefor 'real' news and possibly dangerous talking points; and as a legitimation of company rule through sporting success. Other communal activities which might fit into the same category included Ambulance Brigades, horticultural associations, prize colliery bands (but not trade union bands), and Boys and Girls Brigades; annual trips to the seaside for the entire population of the colliery villages were organized under the companies' auspices.

Paternalism could work on a more individual level too:

> The Bolsover Company, which I say here and now was the Rolls-Royce of the mining industry, definitely, and they were better to us than ever they've known how to be since nationalization. For instance, if one of the village colliers, a workman or an official was ill, it was quite common for the managing director's wife to take him soup, this, that, and the other, you know, tasty bits.[92]

The Butterley Company, like most of the companies which were involved in the new Nottinghamshire field, was very much a paternalistic enterprise. For example, A. Leslie Wright, the uncle of Ollerton Colliery Manager Montagu Wright,[93] was very much an owner of the old school. A member of the Board from 1888 to his death in 1938, Leslie Wright was the Managing Director of the Butterley Company in the 1920s and 1930s when the pit and village of Ollerton were being

[91] See R. P. Dore, *British Factory Japanese Factory* (London, 1973), pp. 65, 204-5.

[92] Interview, J. H. Guest.

[93] Although Montagu Wright (1896-1968) was a member of the family which owned Butterley, he had to serve an apprenticeship in the industry. Having started with the company in 1912, before 1914 he worked in the survey, engineering, and electrical departments. After serving in the forces all through the war, he returned to Butterley and worked at the coal-face for 12 months at Britain Colliery, Derbyshire, then for another year underground at Kirkby, Nottinghamshire, to gain experience of mechanical cutters and conveyors. Even after he obtained his manager's certificate in 1922, he was sent to work as a shotfirer for six months at Hartshay Colliery. He was appointed to Ollerton after answering an advertisement for 'a manager at a new Notts. colliery', and was chosen for the job not by his uncle but by mining agent H. E. Mitton (Mottram and Coote, *Five Generations*, p. 168; M. F. M. Wright, Memoirs).

developed. The historians of the Butterley Company wrote of him,

> Leslie Wright may have been dictatorial in some ways, but he was patriarchally dictatorial. I have seen old family enterprises similarly run in other lands, notably in Portugal. The type of enterprise and the type of man have both gone from this country, and if much that was unpalatable has gone with them, so has much that is good.[94]

The soup provided by the Managing Director's wife at Edwinstowe, like the banquet provided for the sinkers when the coal seam was struck at Ollerton,[95] like the miners' teas at Welbeck Abbey, and the company outings to Skegness,[96] may all have been an example of what R. K. Merton called pseudo-*Gemeinschaft*, 'the feigning of personal concern with the other fellow in order to manipulate him better', 'the mere pretence of common values in order to prevent private interests'.[97] It may have been of a similar nature to the Lincolnshire Harvest Festivals organized by local farmers in the nineteenth century, which Obelkevich described as an attempt to kill discontent by kindness, by 'a prudent and inexpensive investment in social control'.[98]

But it is uncertain to what extent any of these gifts may be seen as deliberate and conscious attempts to reinforce an unequal relationship of patronage and deference between employer and employed. Perhaps the employers themselves believed to some extent at least in the world of harmony and co-operation which Merton and Obelkevich thought had been destroyed by the erosion of 'community' in favour of a class orientation of society. There is little evidence that the Dukeries coal companies, large and efficient though they were, were influenced by theories of scientific management such as those associated with F. W. Taylor.[99] If management was interested in pushing the 'frontier of control' further in

[94] Mottram and Coote, p. 167.

[95] See above, p. 46.

[96] See below, pp. 197–9

[97] R. K. Merton, *Mass Persuasion* (1946), pp. 142, 144.

[98] J. Obelkevich, *Religion and Rural Society: South Lindsey 1825–75* (Oxford, 1976), pp. 60–1.

[99] See Harry Braverman, *Labour and Monopoly Capitalism* (London, 1974), esp. chs. 4–5; L. Urwick and E. Brech, *The Making of Scientific Management* (London, 1951–7, 3 vols.); Daniel Nelson and Stuart Campbell, 'Taylorism versus Welfare Work in American Industry', *Business History Review*, 46 (1972), 1–16.

their favour, and reducing the miners' independence and ability to take important decisions about the work process, this was more a result of what was regarded as common business-sense. In many ways the companies' policy was rooted in a somewhat old-fashioned style of paternalism which strongly resembles that current in British industry in the nineteenth century, before Taylorist theories were ever formulated in the United States. For example, it is clear from the testimony of those who knew him, and from his Memoirs, that Monty Wright of Ollerton was a dedicated and well-intentioned man who genuinely believed that he could protect and further the welfare of all the residents of Ollerton. He professed respect and admiration for his miners, a sentiment which was often reciprocated. The soup purveyed by the Managing Director's wife may have played a role in preserving the social and political status quo in the village, but it is hard to argue that the motives behind her action would have been recognizable to any of the actors at the time.

However, the company's concern with what went on in the village, and their desire to promote harmony between the workforce and the management, had a negative side to it as well. That there existed a feeling that the company dominated the village and its inhabitants was undeniable. For a start, dismissal from the pit meant eviction from a company house: 'If you lost your job at the pit for any reason, you was out of the house like a flash. In fact there was two families turned out, one Saturday afternoon or one Sunday afternoon when it were raining, and the doors were nailed up so they couldn't get in again.'[100] The management was frequently worried by stories of callous evictions following the death of a miner, which did not do their image any good at all. In December 1944, the Revd S. J. Galloway, Vicar of Ollerton, interceded for a widow given notice to quit. Montagu Wright replied that:

> One knows that in a village like Ollerton there are a number of rumours going around, and these are generally of the type which detract from the way the Management are looking after the people, and one never hears much said about the good things that are done.[101]

[100] Interview, C. H. Green, Ollerton.
[101] Butterley H3 224.

Yet the company's heavy hand was not felt only when a man left the pit: 'There was a constant feeling of being company dominated, the presence of a company employed "policeman", complete with pseudo-uniform, being an ever-present figure. I always had the feeling that I was an "extra" in some unscripted Gilbert and Sullivan production.'[102] The only way that one could tell that he was not a real policeman was because of the shortness of his stature and because the initials on his helmet read BC for Butterley Company rather than PC.[103] The uniformed company policemen or watchmen were a common feature in the new Nottinghamshire mining villages before nationalization. Their duties were primarily concerned with the maintenance of the appearance of the village, keeping the lawns tidy and litter off the streets, for example, but they undoubtedly represented and personified the authority of the coal company as well. To take three examples from different villages:

The women, if the kids didn't behave themselves, they used to say, now if you don't behave yourselves, I'll fetch Bobby Healey to you. But Bobby Healey wasn't a big fellow, he wasn't a frightening feller, he wouldn't have struck nobody, but he'd got that much power across at the pit.[104]

If you were out playing and doing nothing to upset anyone and you saw Bobby Copeland coming, you came in your own yard. I used to wait in the kitchen till he'd gone. I wasn't doing anything wrong but that's how we felt about him. And if a lad was misbehaving he'd have a flick with Bobby Copeland's gloves and not a word would be said, and he wouldn't do it again.[105]

JHG Well, he wore a policeman's uniform, and everybody called him Bobby Sand, you know, kids and that, and he used to patrol the village, and I suppose he'd report anybody's garden that were untidy or anything like that, and they'd get a letter from the management.

RJW What did people think of him?

JHG Well, they treated him with respect. Very much respect too.[106]

Bobby Healey of Ollerton was a colourful figure. *The Sphere* recognized the importance of his role in the village in 1932:

[102] Interview, A. E. Corke, Ollerton.
[103] *Idem.*
[104] Interview, C. H. Green, Ollerton.
[105] Interview, Hilda Tagg, Rainworth.
[106] Interview, J. H. Guest, Edwinstowe.

One must not reckon, of course, without Bobbie Healey, the local Solomon, *arbiter elegantiarum*, and public detective, who sees that every household maintains a standard of cleanliness, which at first was strange to the newcomers from Durham and Wales in particular.[107]

Bobbie Healey, once in the 17th Lancers, then a mounted policeman in Egypt (he was one of the four men detailed to arrest Zaghbul) is now 'Works Constable' at Ollerton. But he only wears a uniform on pay day. He has a broad Midlands accent, a high forehead, aquiline nose, policeman's boots, and the judgment of Solomon. He acts as arbitrator, unofficial police court magistrate, and Little Father of the Poor. Employed by Mr. Montagu Wright, as colliery magnate, he settles disputes between neighbours; identifies poachers; takes complaints about leaks in the roofs; pacifies husbands; makes lodgers pay their rent to their landladies; stops small boys from running over their neighbours' lawns; starts the boxing tournaments; takes care that there is no impropriety at the dances; and in fact, acts as the local Solomon.

Under his care there are 820 houses and 3,991 men, women and children. He has been inside every house, and has reported whether it is clean or dirty (at the last census only seventeen came into the latter category); he inspects the gardens, notes whether they are cultivated or not; cracks jokes with the newly marrieds; and informs the police about any men who on rare occasions may have been guilty of theft.[108]

It is scarcely surprising that Healey left a vivid impression in the minds of Ollerton residents of these years, especially those who were children:

RJW Do you remember a man called Bobby Healey?
CB I do indeed, yes!
RJW Who was he?
CB Well, he was the pit bobby, and you were scared to walk across the lawns—it was open-plan lawns in front of the houses—and if he saw you walking across the lawns you felt that he was an ordinary bobby, and you could be fined, imprisoned, whatever—you were as scared of him as an ordinary bobby.
RJW But he wasn't a real policeman?
CB No, he wasn't, we knew he was the pit bobby but we thought he had the same powers as an ordinary bobby. There used to be one or two odd bits of railway line left, sidings which they'd never really been able to use, put the odd trucks and things in there. And when we were doing anything we shouldn't do, it was a joke, 'Oh, Bobby Healey'll be here', and you know you got scared and run away if he was. He was a sort of village ogre.[109]

[107] See above, pp. 45-6 for further examples of this article's view of migrants.
[108] 'A Miner's Life'.
[109] Interview, Cyril Buxton, Forest Town (Ollerton).

The company was determined to maintain the cleanliness and neatness of the model village. Men had to pay 2*d.* a week at Ollerton to have their lawns mown by the company.[110] All the same, Wright complained in May 1937 to the colliery manager W. S. Fletcher that the lawns were in a disgusting state.[111] Men were prohibited from walking on their own front lawns: 'On your own lawn, if you were found walking on them, they used to fine you two shillings, and if you didn't pay it, you could reckon you were for the high jump in those days.'[112] At Rainworth, the new colliery village for Rufford pit:

The Manager came round, and he had these two men from the gardens committee with a sheet of paper jotting down his comments on the gardens. He also had the estates foreman, with his little notebook, jotting down the adverse comments like somebody's hedge weren't right, or somebody's garden was full of weeds, and they were fetched into the office next day and told to get it sorted out or they'd have to go.[113]

It seems clear that the threat of dismissal was thought to hang over men who did not keep their gardens tidy in several villages. At Bilsthorpe:

The only thing that the Stanton Ironworks Company was concerned with, was that you tried to keep your garden tidy, you know. As long as it was tidy they were satisfied, and they held competitions for the best garden of the year, all colliery villages did this. But in this village itself, everybody didn't do their garden, but if their next door neighbour complained, or one or two complained about an untidy garden, automatically the manager had you in the office and said, 'look, if you don't get that ruddy garden tidied up, mate, you're up that pit lane'.[114]

Men were also forbidden to keep dogs as pets in many of the villages, so the popular miner's sport of greyhound or whippet training was outlawed. In July 1935, the manager of Ollerton Colliery stressed that any offender in the matter of dogs 'would be severely dealt with'.[115] When a man named Dowdall was told to get rid of his greyhound, he preferred to give in his notice and leave the village.[116] A relaxation of this

[110] Butterley L1 226. [111] Ibid.
[112] Interview, C. H. Green, Ollerton.
[113] Interview, Hilda Tagg, Rainworth.
[114] Interview, Herbert Tuck, Bilsthorpe.
[115] Butterley L1 226, 15 July 1935. [116] Ibid.

rule did not come until December 1941, when Monty Wright wrote asking if it was right to enforce the rules against greyhounds: 'Do you think we are justified in preventing our workmen from enjoying such sport?'[117]

There can be no doubt that misbehaviour in the village might have consequences in the pit itself:

What one felt very much, you could feel the nuances of things, and yes, one certainly had to toe the line as a miner occupying a company house, otherwise—I really don't know how to put this—otherwise you could not be looked on favourably at work. Remember these were the days of the butty system and in fact I think a good deal of work was done over a bar, or conditioning for work.[118]

Needless to say, the power exerted by the company might b be abused by its officials. On 15 May 1933 a complaint reached Butterley Head Offices from the East Midland District Council of Retail Newsagents, Booksellers, and Stationers. According to this body, the man who let the colliery houses in New Ollerton, Healey, 'lays down a condition that the new tenant must have his newspapers from a Newsagent in Old Ollerton to whom by the way he is related by marriage.'[119] This was damaging the trade of the Council's New Ollerton members. The Butterley mining agent, J. Bircumshaw, asked Monty Wright 'Who is this American racketeer (Mr Healey) they refer to?' but the colliery manager justified the policy with the response that the New Ollerton newsagents' boys walked across the lawns and damaged them!

In many ways colliery company houses were analogous to the tied cottages of Britain's agricultural areas. In his study of rural East Anglia, *The Deferential Worker*, Howard Newby wrote of the hostility of agricultural trade unions to the tied cottage system:

Agricultural trade unions have been implacably hostile to the tied cottage, viewing it as an intolerable restriction on the freedom of agricultural workers. The tied cottage is identified with the exploitation of the farm worker by agricultural employers and as their principal device in maintaining a low-wage regime and consequent poverty on the land. The tied cottage is regarded as the centre-piece of the system which has

[117] Butterley H3 224 H3 224, 18 Dec. 1941, Wright to W. S. Fletcher, manager of Ollerton Colliery.

[118] Interview, A. E. Corke.

[119] Butterley H3 223.

condemned the agricultural worker to poverty and dependence over the last 150 years or more. By restricting the mobility of workers between agriculture and other industries, the tied cottage literally ties the worker to the land. Thus terms like 'serfdom', 'captivity' and 'under the thumb' are part of the armoury of rhetoric brought to describe the dependent situation which ensues. Where an employer is also the landlord, the twin threat of dismissal and eviction is viewed as a concentration of power in the farmer's hands which removes the last vestige of independence and freedom from the already low paid farm worker. This dependency is particularly extreme in remote rural areas where there is little alternative employment or housing. Here the subservience of workers and the payment of low wages is most easily enforced. In many cases, life 'under the boss' can lead to intolerable interference in the private lives of workers and petty restrictions on their leisure activities. The unions also recall that in the past the tied cottage system has been used as a weapon against the spread of unionisation or the threat of industrial action and that personal disagreements with employers over non-farming matters have resulted in genuine hardship for the worker and his family.[120]

Much of this description, *mutatis mutandis*, could be applied to life in the Dukeries mining villages in the 1920s and 1930s. Certainly it reinforces the impression of economic and social influence and power gained by an employer who happens to own his workers' homes. A. K. Giles and W. J. G. Cowie pointed out some of the effects of the tied cottage system:

It is not a simple concept, this insecurity. Being in a tied cottage colours one's attitude to many of life's vicissitudes. Choice of employment becomes confined if it entails house removal as well. Illness is something to be feared more than normally because being laid off for any length of time involves the risk of being dismissed . . . redundancy also poses similar difficulties, should it arise. The occasional news of some fellow-worker's plight under any of these vicissitudes, especially if his employer has been over-hurried in securing possession of his cottage, keeps such fears alive.[121]

It is clear that there were direct political consequences of the companies' influence in the mining villages.[122] Activity on behalf of socialist political groups or the official union, the Nottinghamshire Miners Association, was frowned upon,

[120] Howard Newby, *The Deferential Worker* (London, 1977), pp. 182–3.

[121] A. K. Giles and W. J. G. Cowie, 'Some Social and Economic Aspects of Agricultural Workers' Accommodation', *Journal of Agricultural Economics*, 14 (1960), 155.

[122] See below, chs. 5-6.

and of course it was impossible to book rooms for meetings of this kind. One NMA activist, Sam Booth, was sacked and evicted from his house in Ollerton. Nobody dared to take his family in as lodgers: 'His own son-in-law lived not far away and they daren't take them in for fear of the same thing happening to them.'[123] Similarly Booth was not allowed to collect for the official union on Butterley property. The first time that it was possible to hear a Communist, Gallacher, at an open meeting in Ollerton was in the 1940s, during the wartime coalition.[124] It was recognized that employees living in the colliery villages were more subject to company control than those living elsewhere. As the Stanton Company Colliery Committee minutes put it in 1928: 'The Management cannot be so well in touch with men from outside as with men living in the Colliery Village, and the men from outside are never so amenable to discipline.'[125]

It is clear that one of the reasons why the companies were prepared to invest so much in the provision of housing for their workmen was because they could influence employees living in company houses at the pit gates far more effectively than those living in a more mixed community several miles from their place of work. Neville Hawkins of Harworth suggested that this was the real reason why the colliery villages were built in the first place:

> Why did Barber-Walker build almost 1,100 houses? It wasn't because they were interested in the living conditions of the people, so much as to have a grapple hold on somebody so as he couldn't argue too long or they'd put him out on the street—which they did! Evictions were very commonplace in the thirties, very commonplace.

The economic control of the companies could be devastating. As one interviewee put it:

> This was a prison camp! Absolutely a prison camp, because I'll tell you for why. The Stanton Ironworks Company, they owned the public house, the grocery stores. They owned the post office, the butchery, everything in this village belonged to them. And they could have absolutely paid us out with coupons, at the end of the week in regards to what we would have done. Because the bus services to

[123] Interview, C. H. Green, Ollerton.
[124] Interview, A. E. Corke, Ollerton.
[125] Stanton Company Colliery Committee Minute 477, 8 Oct. 1928.

Nottingham and Mansfield was really nearly non-existent even if you could afford to go, but they were very bad, the services were, and so therefore you were more or less in a settlement, in a commune, and you could have took coupons out instead of money.[126]

Yet the consequences of company control extended far beyond the obvious and direct exercise of power. As J. P. Gaventa pointed out in his study of the case of Appalachian coal mining valleys,[127] the 'third dimension of power', the destruction of any belief in the possibility of militant action against the company, and the establishment of a deferential attitude towards the employers, must also be taken into account.[128] If one rejects the liberal interpretation of the notion of power as portrayed by the school of Robert Dahl[129] and Nelson W. Polsby,[130] which holds that men are always capable of perceiving their own best interests, one can see how the coal companies of the Dukeries could influence the views and values of the inhabitants of the colliery villages. As a result they too could praise the orderliness of a 'model' village, and deny any conflict of interests between themselves and their employers. This indirect use of power by the coal companies complemented their use of coercion against recalcitrant employees, a coercion based on the efficacy of the threat of dismissal in a time of common hardship.

The quiescence of the Dukeries miners is reminiscent in some ways of that exhibited by agricultural workers in rural areas of Britain in the twentieth century. Newby wrote of:

The construction of an elaborate web of paternalistic relationships which not only consigns its subordinates to a dependent and powerless situation but enables them to endorse the system which achieves this . . . the sedative effects of paternalism are of a kind that bring about stability and order and an identification of workers with their 'betters'. Thus while dependence seems to be embedded in the very idea of paternalism, what is of greater importance in producing a stable system

[126] Interview, H. E. C. Tuck, Bilsthorpe.

[127] John P. Gaventa, *Power and Powerlessness: Quiescence and Rebellion in an Appalachian Valley* (Oxford, 1980).

[128] Steven Lukes, *Power* (London, 1974).

[129] Robert Dahl, *A Preface to Democratic Theory* (Chicago, 1956); *idem*, *Who Governs? Democracy and Power in an American City* (New Haven, 1961).

[130] Nelson W. Polsby, *Community Power and Political Theory* (New Haven, 1963).

of traditional authority is not so much material dependence as the dependence of subordinates upon certain definitions of their situation . . . the paternal employer would convey ideologically 'correct' evaluations and moral attitudes, legitimating his own power and defining the worker's role as that of a partner in a co-operative venture. Hence the relationship should be only 'partially authoritative', but rather involve an exchange of 'affectionate tutelage' for 'respectable and grateful deference'. Indeed paternalism involves this dialectic of authority and affection at every level. At one and the same time it may consist of autocracy and obligation, cruelty and kindness, oppression and benevolence, exploitation and protection.[131]

It would be wrong to claim that all the residents of the colliery villages welcomed company policy in its entirety. What, for example, of those who 'voted with their feet' to leave—one can only guess at their attitudes.

In the mining villages of the Dukeries, just as on the farms of East Anglia, can be found both deference and powerless resentment. There are residents of the private company villages who maintain a hatred for the system of tied housing and restriction, embodied in the figures of the colliery manager and the company policeman, the deputies, and the butties. Such men would recognize the traditional ability of the agricultural labourer to tip his cap to the employer while raising two fingers behind his back. However, even if these men cannot be said to exhibit deferential *attitudes*, their *behaviour* was in general still quiescent: little resistance was considered to be possible within the pit or the village. Perhaps 'calculative' deference based on powerlessness can be as potent a force as enthusiastic subscription to the existing order.

On the other hand we can also find those who look back on the days before the Second World War as the golden age of model villages, neat, tidy, and harmonious. These men nurture a great respect for figures like Monty Wright[132] and Bobby Healey of Ollerton, and retain a pride in the success of the colliery sports teams and prize bands. The stability of the society of the colliery villages was not enforced crudely

[131] H. Newby, *The Deferential Worker*, p. 425.

[132] It is interesting that Wright was almost universally known to the miners by his diminutive name. Patrick Joyce (*Work, Society and Politics* (Brighton, 1980), pp. 189-90) identifies this as part of a distinctive Tory paternalist style in the nineteenth century.

and solely by coercion and the threat of dismissal, but by the traditional paternalism and hierarchical authority of the employers. Of course this stability was also underpinned by an acceptance of the 'common-sense' limits to the possibility of radical action and change, and the restricted consciousness of conflict in society, whether local or national. This common sense might be shared without question by employer and miner alike. Even where the personality of the manager or the unsatisfactory and ill-rewarded nature of work was resented by the miner, 'the deeper hegemony of liberal ideas and values remained unshaken'.[133]

We may conclude that the coal companies interpreted the ramifications of their investment in the deep, highly profitable pits of the new Nottinghamshire coalfield with great care and imagination. The colliery villages and all that went on in them were seen as an integral part of the whole venture, and a suitable subject for the concern and intervention of the coal company and its officials. The pit spread its purviews and, as with the nineteenth-century 'factory system', 'work got under the skin of life'.[134] Relationships of employment shaped social relations in the villages. 'Deference', however one defines it, is an outcome of social stratification rather than its source,[135] a symptom of a stable power structure rather than a cause of quiescence.

The new coalfield was seen both as a great new development and a pointer to the future. It was possible in December 1927 for the *New Statesman* to write:

> The continued expansion of the Dukeries coalfield is an assured fact. At the moment its pits are turning out eight or nine million tons of coal. Its reserves are overwhelming. In another generation 50,000 miners will be working and living in the company-owned pits and the company owned villages of eastern Nottinghamshire. The balance of power in the mining industry is shifting inevitably from South Wales and Durham to South Yorkshire and the Dukeries. The foundations of the most important and profitable coalfield in Great Britain are being laid down with an ominous emphasis on the 'butty system', on company unionism, on the company village. A new industrial feudalism is being erected in the

[133] R. Q. Gray, *The Labour Aristocracy in Victorian Edinburgh* (Oxford, 1976), p. 156.

[134] P. Joyce, *Work, Society and Politics*, p. xv.

[135] Ibid., p. 95.

Dukeries side by side with which Trade Unionism can at present find no place. Will it mean the beginning of better things or worse things?[136]

As it happens, despite a vigorous and bitter propaganda campaign in which the Dukeries coal companies figured prominently,[137] twenty years later their investment in the most prosperous of coalfields was overtaken by nationalization of the mines and the end of company control in the villages. But the continued importance of the Dukeries coalfield in the nationalized industry shows that the struggle for unhindered exploitation of the new field by the companies in the inter-war years which had produced the phenomenon of company towns in rural Nottinghamshire, was indeed a struggle in which the stakes were high and the rewards well worth while. The 'industrial feudalism', the company power the *New Statesman* feared, did not prove to represent the future course of the coal industry. But it was not for want of trying on the part of the Dukeries coal companies.

[136] *New Statesman*, 24 Dec. 1927.
[137] Butterley P9 292 'Propaganda', and see below pp. 218–9.

5. Trade Unionism

'United we stand, for union is strength'

'United we stand, for union is strength': such was the motto of the Nottinghamshire Miners Association. Yet the organization of Nottinghamshire miners in the inter-war period was characterized by disunity, schism, the rise of the 'non-political' Industrial Union founded by George Spencer, and (less obvious but just as important) the hidden 'third force' of non-unionism. Several narrative accounts of the Spencer union have been written.[1] Formed in 1926 when the Nottinghamshire miners required a body to co-ordinate their drift back to work at the end of the national coal strike of that year, the Spencer union remained in independent opposition to the 'official' NMA until 1937, when the two unions were combined in the new Nottinghamshire Miners Federated Union, after the Harworth dispute of 1936/7. It is proposed in this chapter not to retell the story of Spencerism, but rather to analyse and explain the differences to be seen in the functions, emphases, and methods of operation of the rival unions by studying central and local branch minute books. Then the phenomenon of the split must be fitted into the general picture of the Dukeries, a new, expanding, prosperous, isolated coalfield suffering the dislocations of migration and of new communities which often took on the aspect of company towns.

'Between masters and men'—the Spencer union

Autonomy was the keynote of the Spencer union, or more properly, the Nottinghamshire Miners Industrial Union. Its

[1] A. R. and C. P. Griffin, 'The Non-Political Trade Union Movement' in Asa Briggs and John Saville eds., *Essays in Labour History*, vol. iii (London, 1977), pp. 133–62; A. R. Griffin, *The Miners of Nottinghamshire* (London, 1962), pp. 203–20; *idem*, *Mining in the East Midlands 1550–1947* (London, 1971), pp. 302–18.

origins lay in Nottinghamshire's reluctance to remain on strike in 1926 for the sake of less prosperous coalfields, a reluctance explained by the wealth and relative harmony of its own industry. Nobody was asking Nottinghamshire miners to take any substantial reductions in wages. The idea of local autonomy is often associated with George Spencer himself. After all, he was still complaining in his retirement in 1944 about the formation of the National Union of Mineworkers.[2] But it was the logic of the miners themselves which dictated the return to work in 1926. None of the Spencer union's branch minute books at any of the new pits sunk in the 1920s has survived, but we do possess the NMIU Branch and Committee Minutes for Rufford Colliery, sunk between 1911 and 1913 and physically the closest pit to the new Dukeries field.[3] These unpublished minutes enable us to trace the history and policy of a strong and typical NMIU branch from its inception in 1926 up to the amalgamation of the Spencer union and the NMA in 1937. The Rufford minutes show clearly that Spencer was not the 'evil genius'[4] behind the Nottinghamshire miners' return to work in the autumn of 1926.

At Rufford Colliery an 'independent pit committee' was elected on 31 October 1926 to represent the men in negotiations concerning the return to work.[5] On 7 November it was proposed and carried 'that we have nothing to do with G. Spencer'.[6] Spencer did not instigate the return to work, but merely intervened in its later stages in an attempt to secure the best terms possible for Nottinghamshire men. By 21 November, the Rufford branch had 'read and unanimously accepted' the Spencer district wages agreement, although the branch minutes were still headed 'Meeting of Rufford Workers'.[7] However it was soon appreciated that some kind of new union *was* needed. At Rufford's fourth meeting on 28 November, 'in his remarks the chairman explained the reason of this meeting which was to fall in line with the rest

[2] A. R. Griffin, *The Miners of Nottinghamshire*, p. 305.
[3] The Rufford minutes were kindly lent to me by Dr Alan Griffin.
[4] But see the views of S. Pollard, *Economics History Review*, 15 (1963), 563–4, and J. E. Williams, *Bulletin of the Society for the Study of Labour History*, 5 (1962), 51.
[5] Rufford Branch Minutes, 31 Oct. 1926.
[6] Ibid., 7 Nov. 1926.
[7] Ibid., 21 Nov. 1926.

of the county, and form an Industrial Union for Nottinghamshire. He then called on Mr G. Spencer.'[8] On 19 December, Mr J. Birkin 'addressed the meeting and in the course of an excellent speech spoke of the rot that had set in in the old NMA and impressed on them the necessity of attending all meetings to stop the same thing occurring in the New Union.' So by the end of 1926 a Spencer-union branch had been formed at Rufford, by a process not of conspiracy but rather 'organized spontaneity'. What were the distinguishing functions and principles of the new union?

Co-operation with the employers was a primary aim of the Industrial Union. Even before its inception, the Rufford men had founded a committee to negotiate 'between masters and men'[9]—terminology borrowed from an age before mass trade unionism. Spencer told the Rufford branch on 28 November 1926 that 'the new organization when formed would tend to promote a better feeling between masters and men.' In February 1927 Spencer official Stan Middup

> spoke on the propaganda of the old Federation[10] and the methods adopted in fermenting [*sic*] trouble in the industrial world and gave his reasons why capital and labour should work amicably together, and hoped that we should all work together to bring the lot of the miner up to a better standing than hitherto and to assure the future of the old miner who had spent all his energies in the industry.[11]

Here we see the spirit of co-operation shading into another typical NMIU doctrine, concern for the aged miner: the Spencer union was not a company or bosses' union without genuine interest in its members' welfare, but did adopt an old-fashioned view of union activity made possible by the prosperity and harmony of the Nottinghamshire coal industry. Owing to these features, there was relatively little feeling of 'us and them' to promote a belief in conflict between employers and workers. Thus we can see how local characteristics can sap any stark view of the coalminer as a 'traditional proletarian' worker with a strong class image of society.[12]

[8] Ibid., 28 Nov. 1926.
[9] Ibid., 31 Oct. 1926.
[10] The NMA was affiliated to the Miners Federation of Gt. Britain.
[11] Rufford Branch Minutes, 13 Feb. 1927.
[12] D. Lockwood, 'Sources of variation in working class images of society', *Sociological Review*, NS 14 (1966), 249-67.

Along with co-operation with the employers, we must consider the anti-socialism, or anti-political unionism, of the NMIU. The leaders of the union took great pains to separate the political and industrial function of a trade union, as the NMIU's very name implies. Clause 2(g) of the Spencer-union Rules stated that 'the union is an industrial organization and no funds of the union shall be used for any political purpose whatsoever.' The Rufford Committee stated as early as November 1926 that 'we adhere to the principle of establishing our own union under the Bolsover Co. and that it be purely non-political.'[13] In March 1928 Mr Winters 'in a very able speech on the policy of A. J. Cook said that in his opinion the solution of the current difficulties in the Trade Union movement was the elimination of politics.'[14] Needless to say, the idea that 'politics' should or even can be excluded from industrial relations depends on those relations already seeming harmonious and satisfactory to all concerned. The Spencer union, like many conservative organizations, considered that only change and socialism merited the description 'political', and that the defence of the status quo was not a political act. Together with opposition to political action went opposition to industrial action. In November 1926, Spencer 'dealt with the late strike and explained the reasons which caused it, terming it in his opinion an experiment in elementary communism. The new organisation when formed . . . would provide old age pensions for all men past work and would eliminate strikes.'[15] We have no example of the NMIU calling a strike in its eleven-year history—it felt that by negotiation and co-operation it could achieve its limited aims.

Non-political the NMIU may have been, but it was vigorous in opposing its rival, the NMA, and competing for members to safeguard its own position. Rule 16(b) of the Spencer union was designed to remove agitators from the ranks: 'Communists and members of the Minority, or any similar revolutionary movement shall not be eligible for membership or entitled to benefit and the Council of the Union reserves to

[13] Rufford Branch Minutes, 3 Nov. 1926.
[14] Ibid., 11 Mar. 1928.
[15] Ibid., 28 Nov. 1926.

itself the right to expel any member found to be in membership with any such movement or advocating its policy.'[16] There were many attempts to enforce an NMIU closed shop in various pits. At Rufford the branch committee was constantly concerned with this; let us take a few examples:

Moved and Sec. that Sec. write for an interview to Mr. Wheatley re it being a condition of employment that all men set on in future be asked to join our union.[17]

That a letter of protest be sent up to the management against those men who have gone into the new work and who are not in the union; S. Davis, J. Holden, J. Phillips, E. Marshall, W. Robinson.[18]

List of non-union men got out to be sent to the management.[19]

Concerned with non-union men in the No. 2 district and the question of closer cooperation between the management and the Industrial Union.[20]

The question of Union men being removed and non-union men being allowed to stay in decent places.[21]

Complaints from our members being sent into Stalls as Daymen, particularly working for non-union men.[22]

Some attempts were made to win over NMA members by offering them favourable terms: 'That we accept members of the NMA who are clear on their books as members of this organisation without entrance fee.'[23] More often, though, the NMIU could rely on the policy of the employers to prevent the NMA from operating on pit premises or collecting their dues—the Bolsover Company, for example, operated an effective closed shop at Thoresby and at many pits the NMA branches were destroyed.

Another main difference between the two unions concerned their structure. The NMIU adopted an authoritarian constitution: 'the holding of branch meetings shall be left to the discretion of the Branch Committee.'[24] At Rufford, after the early days of independent 'workers' meetings', the NMIU Branch Committee came to meet at least monthly, while general meetings of the men were reduced to two a year, at

[16] NMIU Rules.
[17] Rufford Branch Minutes 30 Dec. 1926.
[18] Ibid., 4 Dec. 1927.
[19] Ibid., 21 Nov. 1928.
[20] Ibid., 14 July 1929.
[21] Ibid., 9 Feb. 1930.
[22] Ibid., 8 Feb. 1931.
[23] Ibid., 5 Jan. 1930.
[24] NMIU Rule 22.

which elections to the committee took place; these were carefully organized, with speakers brought in specially from the NMIU Council to boost morale. Power was maintained among a small number of 'reliable' officials to reduce the possibility of militant activism from below. Difficult matters were referred by the branch committees to the central NMIU organization, and often to Spencer himself. The idea of the sovereignty of the branch general meeting as enshrined in the NMA constitution disappeared, to be replaced by a hierarchic bureaucracy.

This structure may be partially explained when we turn to consider the more positive functions of the NMIU. The Rufford branch, for example, was particularly concerned with the provision of pensions for aged miners. In November 1928 they called for 'an Old Age Pension fund to be advocated from the Welfare Fund.'[25] In August 1929 the branch was moved to protest strongly against the administration of the welfare scheme—more should be spent on pensions.[26] In February 1933 Rufford was still complaining about the funds of the welfare levy not being turned over to the Old Age Pensioners.[27] A second sphere of action concerned compensation payments. Spencer was justly renowned for his handling of compensation cases; in 1930 alone he won over £10,000 for Industrial Union members. The minutes of the Rufford branch show great concern for the welfare of individual members of the union, and also for the representation of the men in everyday business.

The most frequent method of dealing with problems as they arose was to send a deputation to the management. Between 1927 and 1936 deputations were sent concerning price lists,[28] overtime,[29] bonus percentages,[30] coal allowances,[31] non-unionism,[32] falls of roof,[33] timber shortage,[34] half a dozen problems ranging from rates for machine-cut coal to a shortage of nails,[35] the slack question,[36] to ask the

[25] Rufford Branch Minutes, 28 Oct. 1928.
[26] Ibid., 11 Aug. 1929.
[27] Ibid., Annual Meeting, 12 Feb. 1933.
[28] Ibid., 25 Nov. 1928.
[29] Ibid., 3 May 1930.
[30] Ibid., 21 Sept. 1927.
[31] Ibid., 1 Apr. 1928.
[32] Ibid., 30 Sept. 1928, 11 Aug. 1929, 21 Sept. 1930.
[33] Ibid., 11 Nov. 1928.
[34] Ibid., 17 Mar. 1929.
[35] Ibid., 8 Apr. 1929.
[36] Ibid., 1 Nov. 1931.

manager if he would 'take us a little more into his confidence in the matter of issuing notices',[37] small coal,[38] notice given to 95 members,[39] complaints about the back yards in the Model Village,[40] two men being engaged to take charge of the cycle sheds at the colliery,[41] and shortage of timber.[42]

If this procedure failed, the next step was usually to call in Spencer himself, to talk personally to the management. Spencer also made frequent visits to collieries to explain the price lists he had negotiated with the owners. Other fields of activity included pit inspections, usually after each serious accident, and attempts to protect NMIU members from lay-offs, and to secure reinstatement. Special committee meetings were called even for individual cases, and officials of the union appeared in court to support members in compensation cases or appeals against wrongful dismissal. The branch committees dealt with every matter that might crop up, from shortages of tubs to 'the way in which some of our members are treated by certain deputies'.[43]

It would be wrong to condemn the Spencer union simply as a 'bosses' union'. Within the framework of their ideology, itself a result of the peculiar economic conditions of Nottinghamshire, crystallized by the circumstances of the end of the coal strike of 1926, the union officials in general worked sincerely and vigorously in the interests of their members. If we move on to consider the NMA in this period, we shall see where the differences lay in the functions and approaches of the two unions; these may be resolved into a matter of emphasis rather than into a contrast between legitimate trade unionism and company unionism.

'The finest asset a worker can have': the Nottinghamshire Miners Association

First of all, the political philosophy of the NMA shows distinct differences from that of the Spencer union:

[37] Ibid., 15 May 1932.
[38] Ibid., 25 Sept. 1932.
[39] Ibid., 30 Apr. 1933.
[40] Ibid., 31 Dec. 1933.
[41] Ibid., 12 May 1934.
[42] Ibid., 10 May 1935.
[43] Ibid., 29 Jan. 1928.

Membership of a trade union, still the finest asset a worker can have: it is the only source through which he can make known his grievances, improve his wages, add weight to his social status, claim equality of freedom, and give protection to his family. It is difficult to understand any miner having a leaning towards a non-political organisation, when his very life depends upon politics.[44]

Politics in a wider sense, national and even international, played a role in the NMA, unlike the Spencer union. In 1934 the NMA sent best wishes to the Welsh miners engaged in their fight against non-political unionism.[45] In 1935 Aneurin Bevan was invited to speak at the Kirkby-in-Ashfield May Day.[46] In 1936 a resolution was passed by the NMA Council supporting Communist affiliation to the Labour party.[47] Candidates were sponsored for local authority elections: in the 1930s the NMA backed candidates for the following bodies:

Borough Councils: Nottingham, Mansfield.
Urban District Councils: Beeston, Warsop, Mansfield Woodhouse, Kirkby-in-Ashfield, Hucknall, Arnold.
Rural District Councils: Basford, Colwick.
Board of Guardians: Mansfield.
Nottinghamshire County Council: Mansfield, Mansfield Woodhouse, Gedling.

The NMA was affiliated to the following constituency Labour parties: Broxtowe, Mansfield, Nottingham, Rushcliffe, Clay Cross, Newark, Bassetlaw, and Belper. Grants were made towards the cost of fighting the general elctions of the 1920s and 1930s.[48]

The finances of the NMA were severely affected by the debt incurred during the 1926 coal strike, as Table 23 shows. A more general picture of the income and expenditure of the NMA compared with that of the NMIU is given by Table 24.

One of the factors which led to the gradual reduction of the NMA to a militant rump was the way that employers placed obstacles in the way of the operation of the older

[44] NMA Minute Book, 1935.
[45] Ibid., 1934.
[46] Ibid., 1935.
[47] Ibid., 1936.
[48] Ibid., *passim*.

Table 23. *NMA expenditure 1926*[a]

	pension	out-of-work pay	strike pay
	£. s. d.	£. s. d.	£. s. d.
Jan.	1114 10 0	88 10 6	
Feb.	1084 12 0	420 11 6	
Mar.	1064 18 0	529 6 3	
Apr.	1070 2 0	44 12 0	
May	824 14 0	79 9 3	23600 18 10[b]
June	106 15 0	4 7 6	7008 19 6[c]
July	3 10 0		12433 18 2
Aug.	12 9 6		3512 0 0
Sept.			10045 2 1
Oct.	12 12 0		6639 15 11
Nov.	15 18 0		1798 10 1
Dec.	11 15 0		919 2 10

[a] NMA Minute Book, 1926.
[b] Derbyshire Miners Association donation.
[c] Miners Federation of Great Britain donation.

Table 24. *NMA and NMIU membership and income 1927–37*[a]

	NMIU		NMA	
	members	income (£)	members	income (£)
1927	12,583	18,182	16,624	18,472
1928	15,086	22,792	13,950	10,303
1929	10,774	19,016	13,934	9,527
1931	11,182	16,541	12,295	10,222
1932	12,945	16,485	11,065	9,369
1933	14,524	19,622	10,125	7,840
1934	15,046	18,121	8,500	7,493
1935	16,644	19,169	9,869	8,185
1936	16,948	19,946	11,146	16,669
1937	17,179		9,700	

[a] Annual Reports of the Chief Registrar of Friendly Societies, quoted by Griffin and Griffin in Briggs and Saville (eds.), *Essays in Labour History*, vol. iii, p. 157.

Table 25. *Union membership in Nottinghamshire, 1927-37*[a]

	numbers employed	% NMA	% NMIU
1927	57,955	28.7	22.1
1928	52,114	26.8	28.9
1929	52,702	26.4	20.4
1930	52,393	25.7	21.9
1931	51,307	24.0	21.8
1932	49,499	22.4	26.2
1933	46,909	21.6	31.0
1934	46,852	18.1	32.1
1935	43,923	21.5	36.2
1936	45,538	24.5	37.2
1937	45,579	21.3	37.7

[a] Griffin and Griffin, 159.

Table 26. *Percentage of NMA members 1934*[a]

Kirkby Summit	69.8
Cinderhill	61.2
New Selston	55.2
Silverhill	47.1
New London	43.4
Hucknall no. 1	40.9
Clifton	39.0
Pye Hill	35.9
Watnall	35.3
Moor Green	34.8
Newstead	32.7
Bestwood	29.5
Linby	28.5
Lodge	28.4
Bentinck	20.5
Hucknall no. 2	20.3
Radford	19.1
Annesley	18.9
Wollaton	16.8
Langton	16.3
Bulwell	15.2

Table 26 (*cont.*)

Sherwood	15.0
Mansfield	11.3
*Blidworth	9.1
*Clipstone	8.5
New Hucknall	3.8
*Bilsthorpe	2.7
Gedling	2.0
*Ollerton	1.6
*Harworth	0.3
*Thoresby	0.2

[a] Compiled from the Board of Trade's *Annual List of Mines* and NMA Minute Book, 1934.

* Pits sunk after 1918.

union, refusing to recognize its representatives or to allow the NMA to collect subscriptions at the pit. In some cases, a 'closed shop' was enacted against the NMA and in favour of the Spencer union. In some areas the NMA was almost extinguished, while in others a flourishing branch was maintained. In the western half of the Nottinghamshire coalfield, traditions had been built up which were hard to destroy, and the influx of South Welsh miners to Kirkby after the First World War certainly contributed to the establishment of a large and active branch there. To a large extent the success or failure of the NMA was dependent upon the location of the pit and the policy of the employers.

'The owners are victorious . . .':[49] Trade unionism in the new pits

The new coalfield which opened up in eastern Nottinghamshire in the 1920s formed the bedrock of Spencer union support. In many of the new pits, the NMA never became established, or quickly died. By 1935 there were three NMA members at Thoresby (workforce 1,379), six at Harworth

[49] *New Statesman*, 24 Dec. 1927.

(2,355), twenty-five at Ollerton (1,591), and forty at Bilsthorpe (1,469).[50] The NMIU suffered no such problems of lack of membership and lack of funds in the new pits of the Dukeries coalfield. In 1936 there were 765 Spencer Union members at Rufford, 691 at Clipstone, 702 at Thoresby, and 550 at Ollerton.[51]

No stoppage arising from a dispute originating in the Sherwood Forest pits occurred before nationalization,[52] a feature perceived by residents as well as by outside observers:

FT I think there wasn't even a dispute. I can only ever remember one day's strike at Thoresby, and that was him from here, went mad that day.

GC Oh, that was the late 1960s, we had one day.

FT A one day's strike, that's all we ever had at Thoresby.

RJW What about the General Strike of 1926?

GC Oh, it was only in the sinking processes then, it had only just started taking the sods off in 1926.

FT Yes, they were only a few yards down, and contractors were doing that, so it didn't alter.[53]

Ollerton Colliery was idle only for a month,[54] and since Ollerton was a productive pit by that time, having struck coal in 1925, this was a little more serious. Lord Savile's agent, John Baker, wrote to the Butterley Company on 6 August 1926:

Lord Savile very much appreciates the fact that you have been able during the course of the trouble in the Mining Industry, to quietly progress with the development work at Ollerton and his Lordship quite understands that it is desirable that this should be proceeded with, without attracting attention by publishing what is being done in the newspapers.[55]

At Blidworth, the commitment to the miners' strike and the General Strike was minimal. On 4 June 1926, under the heading 'Back to Work—Blidworth Men set an Example', the *Mansfield Reporter* wrote: 'Blidworth has loomed large in the

[50] NMA Minute Book 1935, *Board of Trade List of Mines.* See above, Table 28.

[51] Butterley T1/6 345 'NMIU 1926–45' (*sic*); Bolsover Company Managers' Annual Reports.

[52] For Harworth, see below, pp. 129–30

[53] Interview, Frank Tyndall, George Cocker, Edwinstowe.

[54] M. F. M. Wright, Memoirs, p. 34.

[55] Butterley F1/1 202, 'John Baker 1924–36'.

press of the country this week through a number of the men returning to work at the new colliery. It will be remembered that the Blidworth men were the last to come out in the district, and now a number of them have set an example to their fellows by returning.'

Why was the Spencer union and the spirit of industrial co-operation so strong in the new coalfield? The first cause that we should consider is that the coal companies, so powerful in the 'company villages' of the Dukeries, favoured the NMIU in every way possible. Often only the Spencerites were allowed to collect levies in the pit yard or were recognized as representing the miners by the management. Indeed, many companies, including Barber-Walker (Harworth), Stanton (Bilsthorpe), Bolsover (Thoresby, Clipstone, and Rufford) and Newstead (Blidworth), helped the NMIU greatly by agreeing to stop the men's weekly contributions to the Spencer union automatically at the pit office along with the other deductions unless the miners 'contracted out'. The Butterley Company was finally persuaded to adopt this policy at a meeting of the Notts. Coal-Owners Association in the summer of 1936.[56] Despite their reluctance to encourage the NMIU in this way, Butterley were as hostile to the NMA as other companies in the new field, as evidence from the Butterley records clearly indicates.

Butterley's mining agent H. E. Mitton wrote to Spencer on 27 January 1927 to tell him 'you have my whole-hearted support'.[57] In May 1930 NMA official W. Carter complained to the deputy mining agent J. Bircumshaw that Montagu Wright, the manager of Ollerton Colliery, refused to answer his letters pleading on behalf of the older workmen when Ollerton Colliery laid off a large number of men.[58] Butterley had never encouraged trade unionism on their premises. In 1925 they had refused the NMA the use of a kiosk in the pit yard; F. B. Varley had warned Bircumshaw that 'by refusing moderate requests the French aristocrats paved the way for the Revolution and lost their heads. Conservative governments are wiser. Only employers are foolish.' Bircumshaw replied that the union's box had been filled with 'Communistic Russian

[56] Butterley T1/6 345. [57] Ibid. [58] Ibid.

rubbish'.[59] In general, Butterley preferred that their employees should not be a member of any union.

Other companies, however, were more inclined to see in the Spencer union a possible way of ensuring continuity of production and dispelling strife, and positively encouraged their men to join the NMIU. At Bilsthorpe, Herbert Tuck felt that a Spencer-union closed shop was in operation, and it is quite understandable how here the NMA seemed scarcely to exist:

RJW Now there were of course at that time two unions in Nottinghamshire, the Spencer union and the older union—

HT There were only one! The Spencer's union.

RJW In Bilsthorpe?

HT That were all over, there was only one. The other one was outlawed!

RJW And so you wouldn't find any members of the old union in Bilsthorpe?

HT Oh, you would. I was in it, but we had to take our money secretly to collectors. You couldn't set up a place in the pit yard to pay, it wasn't allowed.

RJW The company only recognized the Spencer's union?

HT Oh, they formed it! They formed it. What happened was that the Miners Federation was finished completely in Nottinghamshire when George Spencer, in collaboration with the coal-owners, set up this union, called Spencer's union. . . . But then of course it was quite easy; the coal-owners then, before you got a job, you signed to be in that union, and if you didn't sign you didn't get a job! Of course another attraction was, it was only 6*d* a week when the others were paying a shilling, you know, but the only road you could be in if you worked in a Notts. pit, you had to be in two unions. Now that gets back to your two unions. You had to let it be seen that you paid your 6*d* a week to Spencer's union, for your job, and then if you had spirit enough, and principle enough, you also paid into the Federation, to a collector, and in my period of time, in the early days, we had to send it to Blidworth, I think it was.

RJW Is that what you did?

HT Yes, but there was only about half a dozen that did it, for the simple reason that unions had no power, no power at all. Even the trade union leaders, you know, when they were chosen; but all the men were under a cloud in regards to this, that you daren't do a deal of complaining to the manager or else you were creating the character of a nuisance, and they got shut of you! The first

[59] Ibid., 344, 'Notts Miners Association'.

> thing that you did wrong in the pit, that they could hang on you, and make it stick, you were gone![60]

An interviewee less hostile to the coal-owners, J. H. Guest, who was for many years a Bolsover Company deputy, also stated that NMIU membership was compulsory: 'When I went there (Thoresby Colliery), I was informed that I'd have to be in Spencer's union, which I already was.'[61] Actually it is clear from the membership figures that many men at Thoresby and Bilsthorpe, as in most Nottinghamshire pits at this time, belonged to no union at all; but the beliefs of the interviewees suggest that certain companies did exert some kind of pressure to join the NMIU.

This has some basis in the business records of the colliery companies. In September 1927 the Stanton Company's General Collieries General Manager N. D. Todd[62] spoke of the policy at the new pit of Bilsthorpe:

> Mr. Todd reported that he had had an interview with Mr. Spencer, the Secretary of the Miners Industrial Union, who had promised to furnish him with the names of men requiring work, and steps were being taken to employ only men who are members of that Union, if they belonged to any Union.[63]

Certainly an attempt to found an NMA branch at Bilsthorpe in 1929 was vigorously resisted by the Stanton Company.[64] Indeed a question was put in the House of Commons to the Minister of Mines on the subject of whether men were being

[60] Interview, H. Tuck, Bilsthorpe.

[61] Interview, J. H. Guest, Edwinstowe.

[62] N. D. Todd, born 11 June 1884 at Hetton-le-Hole, County Durham; ed. King William's College, Isle of Man. Apprenticed 1902 to Blackwell Colliery Company. Studied mining at Sheffield Technical College, 1st Class Diploma. Manager of Alfreton and Shirland Collieries of Blackwell Company. Agent and general manager of Blackwell Colliery Co. Agent of Stanton Co. 1917–25, collieries general manager 1917–25. Active sportsman, a season for Derbyshire County Cricket Club. 'He fostered sporting facilities and competition of all kinds for the Colliery's employees.' Active participant in the work of the St. John's Ambulance. Officer (Brother) in the St. John's order and Northern Area Commissioner for Notts. On Blackwell and Sutton-in-Ashfield local councils. Left Stanton 31 Dec. 1946 on nationalization to become Areas nos. 2 and 3 general manager for the NCB before retirement in 1952. Died 12 May 1959. (Stanton Company History, vol. 8, 'One Hundred Names to Remember'.)

[63] Stanton Company Colliery Committee Minute 345, 12 Sept. 1927.

[64] See below, p. 183.

dismissed at Bilsthorpe because they were members of the Nottinghamshire Miners Association. On 16 July 1929 Seymour Cocks (Labour, Broxtowe) asked the Secretary for Mines

> whether he is aware that immediately on the formation on 16th June of a branch of the Notts Miners Association at Bilsthorpe Colliery, the president, treasurer, and two members of the committee of the branch were dismissed from the colliery, whilst a delegate of the branch was informed by an official of the colliery that he would be discharged if he continued to have anything to do with the NMA; and whether he intends to institute an inquiry on this matter?

Mr Turner replied that 'Inquiries are being made but are not yet completed. I should be glad, therefore, if my hon. Friend would be good enough to repeat his questions next week.' When Cocks repeated his question on 25 July, Turner replied

> With regard to the first matter I am informed that the four men referred to were presumably four out of 15 men who were dismissed as a result of a certain face being closed. The total number so displaced was 70, of whom 55 were absorbed in other parts of the pit and 15 could not be so absorbed. I am assured that the selection of none of the 15 was due to their trade union membership or activity. I am further assured that the policy of the Nottinghamshire Coal Owners Association and of the owners of Bilsthorpe Colliery remains as stated to representatives of the Trade Union Congress last year, namely 'that they make no inquiry or discrimination with regard to a man's employment as to his membership of any Trade Union.' With regard to the second matter, I am assured that if any such ultimatum was given (and this is denied), it was contrary to the policy and definite instructions of the owners and manager.[65]

This answer is contradicted not only by the oral testimony of many who lived in the Dukeries coalfield in the 1920s and 1930s but by the business records of the coal companies operating in the new fields. In fact it is clear that the co-operative spirit of Spencerism was very useful to the employers, who encouraged the NMIU at the expense of the NMA in various ways. Gascoyne, the first Treasurer of the NMIU, wrote to H. E. Mitton of the Butterley Company on 10 January 1927 asking if they could co-operate to 'counteract the activities of A. J. Cook in the coalfield'.[66] In February

[65] *Hansard*, 5th series, vol. 230 (July 15–26, 1929), col. 262, 16 July; col. 1533, 25 July.
[66] Butterley T1/6 345.

of the same year the same official wrote to Butterley's deputy mining agent J. Bircumshaw, saying that 'the red men can only be kept in Check by joint action'. In the turbulent year of 1930 Gascoyne warned that 'the Reds are fast gaining the ascendancy at Ollerton'.[67] It is scarcely surprising that, given this bitter inter-union strife, if the Ollerton Colliery NMIU checkweighmen Sam Brown and Aaron Jenkins spoke on the radio 'for the men' in 1936, Bircumshaw could write: 'I think this a good Opportunity for a little propaganda on behalf of the colliery owners.'[68]

Besides encouraging the Spencer union, the companies could clearly exert direct force against the NMA in their colliery villages. When in 1929 to 1930 the NMA officials in Ollerton were 'very busy now trying to strengthen the branch they have started there',[69] the Butterley Company responded by sacking the NMA secretary Sam Booth, forcing him to leave his house and the village;[70] he was replaced by Spencer-union man Aaron Jenkins as checkweighman in June 1930. Economic power could be used very effectively in an age of high unemployment and poor unemployment benefits:

RJW Now you've said that there were a few of you who were also members of the other union.[71] Can you remember the names of any of them apart from yourself? What kind of people were they?

HT Oh, militant! There were a chap called Albert Henson from Mansfield Woodhouse, and then I can't really tell you. There were a Taffy Williams who wasn't here very long, came from Wales, and then as soon as he started asserting himself, you know, he got sacked. There was a lot of that type you know, because if you believed in something, you see, that was the trouble; that when we believed in something we used to assert our authority more in claiming our rights and so therefore you got sorted out if you wasn't very careful.[72]

However, the coal companies did not have to use their power of dismissal constantly; there were more regular forces working in favour of quiescence and against militancy all the time. The butty system, prevalent in the new Nottinghamshire coalfield as in the old, placed men under sub-contractors or

[67] Ibid., 10 Apr. 1930.
[68] Butterley L5/3 231.
[69] Butterley T1/6 344.
[70] Interviews, C. H. Green, Ollerton; H. Booth, Langold.
[71] i.e. the NMA. See above, p. 121.
[72] Interview, H. Tuck, Bilsthorpe.

chargemen. This was a well-known curb on industrial action as well as a method of 'divide and rule':

BB Your father would earn more than my father earned in a week because he was a butty—we called them butties—because I mean they were the men who contracted for the work, and the men who actually *did* the work, the butty took the money and paid them out what he wanted to pay them, that was the old butty system in the pits!

RJW Was there any feeling between the butties and the other men?

BB It was a cause of bad feeling quite a lot in the village. At Edwinstowe it was, anyway, I have to tell you that, listening to my father talk about this, in the old Colliery Institute at Edwinstowe they all had to sit on one set of seats, the butties did, and the men sat on the other side.

CB Class distinction.

BB Yes, definitely! Definitely! The butties' children always had more than we did.[73]

The management would pay the butty directly and he would distribute it as he wished amongst the daymen in his team. This was often seen as inequitable:

I can remember, it was somebody called Poyser, this Poyser were actually doing some work and he broke his leg; the butty broke his leg and he was in hospital, and I can remember his wife coming in a small Austin 7, '30s Austin 7, and she came, drew the money from the pit and paid through the window of her Austin 7 the men who'd actually done the work![74]

The importance of the butty system in aiding industrial peace and the Spencer union in the new Dukeries coalfield should not be underestimated. R. Goffee pointed to the importation of the butty system to Kent as one of the most significant causes of dissatisfaction in that new coalfield.[75]

When considering reasons for the political quiescence of working-class people in industrial Britain, and their failure to develop a coherent or revolutionary class consciousness, the notion of labour sectionalism has on occasion been called into play.[76] There is no doubt that the classic butty system

[73] Interview, Cyril and Barbara Buxton, Forest Town.

[74] Interview, Neville Hawkins, Harworth.

[75] R. E. Goffee, 'The Butty System and the Kent Coalfield', *Bulletin of the Society for the Study of Labour History*, 34 (1977), 41–55.

[76] R. Q. Gray, *The Labour Aristocracy in Victorian Edinburgh* (Oxford, 1976); Geoffrey Crossick, *An Artisan Elite in Kentish London 1840–1880*

of the Midlands coalfield introduced more than a little divisiveness, as sub-contractors interposed themselves between owners and face-workers. However, despite the residential segregation of the model villages, it is doubtful if the butties of the inter-war Dukeries coalfield could be described in any way as 'labour aristocrats', and it is also questionable if their role was one of the most pressing causes of the lack of resistance and organization of the miners. By the 1920s and 1930s, butties did not represent a separate craft with its own skills and extensive social and industrial institutions. They did not demonstrate a notable commitment to mid-nineteenth-century virtues of thrift, independence, and respectability. Finally, it seems that by the time the Dukeries field reached full production, the butty system was fading away—certainly it retained little more than a vestige of its former importance. The butty became just a kind of foreman, or perhaps a member of lower managerial supervisory staff. Most of the miners interviewed regarded the butty as a company man, rather than a fellow employee; the classic contradictory commitments of the foreman to the worker and the firm seemed to have been resolved in favour of the latter.[77] The employers appear to have been determined to take the important decisions themselves rather than to leave the butty with a significant degree of independence, as the modernity and high level of mechanization of the new pits perhaps dictated.

The truncated nature of the butty system is perhaps revealed by the fact that few miners can now remember exactly when or how it disappeared. Indeed, comparisons with other parts of Nottinghamshire and elsewhere suggest that the political characteristics of the new coalfield may be explained not so much by buttyism or even by *any* unusual characteristics of the work process in the pit itself, but by

(London, 1978); John Foster, *Class Struggle and the Industrial Revolution* (London, 1974); A. E. Musson, 'Class struggle and the labour aristocracy', *Social History* (1976), pp. 335-56; H. F. Moorhouse, 'The Marxist Theory of the Labour Aristocracy', *Social History*, 3 (1978), 61-82; Alasdair Reid, 'Politics and economics in the formation of the British working class: a response to H. F. Moorhouse', *Social History*, 3 (1978), 347-61.

[77] Joseph Melling, 'Non-commissioned Officers: British employers and their supervisory workers 1888-1920', *Social History*, 5 (1980), 183-222.

the power of the economic monopoly of the companies in these isolated and vulnerable communities. Relationships between managers, foremen, and men in the mine itself seem to have been similar throughout the Nottinghamshire coalfield. The degree of authority transferred from the pit to the village in the Dukeries, however, and the fact that the local labour market had been cornered by the coal-owners, were points of contrast.

The constant threat of unemployment may have been reinforced by the presence in the pit yard of men seeking work. J. H. Guest of Edwinstowe, a Bolsover Company deputy, described them as 'pigeons': 'I used to say as we came out of the pit at the end of the day's shift there'd probably—I used to call them pigeons—there'd be twenty pigeons sat on the railings waiting for a job. So you had to co-operate with the management, or they weren't long at getting you out.'[78] The men seeking employment were not kept outside the pit gates but actually allowed in the pit yard, where they acted as living testimony as to how beholden the miners were to the companies for their livelihood. At Edwinstowe, there was

> a bloke called Tagg, a brilliant bloke, a brilliant colliery man, and the people used to congregate around the two steps up into the offices, and he used to come to the door, and he'd point 'you, you, and you' and those were the people he were going to set on. The rest of them—he just turned his back and walked in, said nothing to them at all, he just picked these people out.[79]

It is understandable that in such circumstances as these the company rarely found it necessary to exercise their full powers in dealing with militants and trouble-makers. Such opinions and actions as were likely to endanger continuity of production were nipped in the bud by the unfavourable market for labour, besides the coal-owners' domination of the villages.

It was recognized at the time that the Dukeries were a black spot for trade unionism. The *New Statesman* summed the situation up in 1927:

> The new coalfield has started life with an evil reputation. It is a blackleg

[78] Interview, J. H. Guest, Edwinstowe.
[79] Interview, F. Tyndall, Edwinstowe.

district envied for its prosperity, despised for its weakness. Through the whole of the bitter lockout of 1926 the wheels in the Dukeries never stopped winding; in the heart of the stoppage the Staveley pits were turning 10,000 tons of coal a day. It was the secret meeting of the coal owners with the 'breakaway' men at Edwinstowe Hall,[80] in the centre of the Dukeries, that smashed the resistance of the Notts Miners Association and created the germ of Spencer's non-political Unionism. The men would not fight, the stimulus of poverty which hardened the other coalfields did not act here. The South Welshmen, from whom so much was hoped, who were expected to infuse and bind the fighting spirit in the Dukeries, stood aloof, never amalgamated, never trusted, never led, fell in with the rest.

The men had their excuses. No appeal could be made to their Trade Union, since in the Dukeries as in every other part of the Notts coalfield, the force and efficiency of the Notts' Miners Association had become a thing to mock at. No Trade Unionist appeal could be made to their loyalty to their county, since they were all strangers, both to the Association and to one another.

The villages, judged by mining standards, are isolated. It is not easy for the organisers to obtain or maintain contact with them. They are owned body and soul by the colliery owners—houses and shops and pubs and picture palaces. The agitator is suspected and marked. Standing by the Federation a year ago meant dismissal. Dismissal meant the loss of a well-paid job, of a clean, comfortable house, and emigration back to the 'short time', the minimum and rotten squalid life of the older fields. At best the fighter could visualise nothing but years in a poor place in a poor 'stall'.[81] There was so little to gain, nothing but the consolation of a spiritual loyalty; and so much to lose.

The owners are victorious. The Notts' Miners Association has, in effect, disappeared from the Dukeries. The men who fought are still hanging around the streets of Mansfield and the neighbouring villages, collecting for the Union, opening back-streets shops to live, looking for jobs of any kind. The pits are closed to these men, men with twenty and thirty years of pit experience, men to whom the pit is a second home, men with three generations of mining blood in their veins. There is no room for them in the Dukeries pits.[82]

There was no challenge from the NMA in the new pits until the mid-1930s. In the NMA Minute Book for 1935 we find the following impassioned plea:

The year has brought to the NMA a very healthy branch at Gedling

[80] The headquarters of the Bolsover Company in Edwinstowe.

[81] The Rufford Branch Minutes of the NMIU often stress the Spencerites' attempts to persuade the management to give their rivals' members the worst workplaces (stalls). See above, p. 112. Here too the butty system, with its capacity for unfavourable allocation of work, could be used against the NMA members.

[82] *New Statesman*, 24 Dec. 1927.

Colliery, the miners caught the determination of our branch at Sherwood, which records almost a doubled membership. Our Harworth branch is growing week by week; the Financial Secretary and the President are ploughing the fields at Ollerton and Bilsthorpe, their efforts are beginning to ripen into added membership; still there is more ground to be broken, more members to be enrolled, improvements for the organisation to be brought into effect . . . let our motto for 1936 be 'all hands to the plough' enrolling members for the NMA.[83]

It was the refoundation of the Harworth branch that led to the bitter dispute of 1936 to 1937 and the fusion of the NMA and NMIU in the Nottinghamshire Miners Federated Union.

There is no need for a further detailed narrative of the Harworth dispute.[84] However, we should attempt to explain how one of the apparently quiescent new company villages of Nottinghamshire came to take its place in the mythology of the labour movement and the Left in Britain—Harworth is almost as well known as a symbol of successful struggle as events in East London's Cable Street, also in the mid-1930s. First, it should be made clear that the re-formation of the NMA branch at Harworth Colliery was almost entirely inspired by external activity, encouragement, and infiltration. In 1935, before the campaign started, there had been only six members of the NMA at a pit with a workforce of nearly two thousand. Mick Kane and many of the other labour leaders in the dispute had come in from outside at a very recent date. Nor is it surprising that Harworth men should join in the strike when given a lead. As in the case of the Second World War,[85] external influences released impulses long restrained by the powerlessness which was the normal state of affairs for the miners of the new villages. Finally, it should not be forgotten that Harworth was on the border with Yorkshire; Spencerism remained predominantly a Nottinghamshire phenomenon, despite the apparently similar economic and social conditions pertaining in the neighbouring South Yorkshire field.

[83] NMA Minute Book, 1935.

[84] See Griffin and Griffin in Briggs and Saville; Griffin, *Miners of Nottinghamshire*, pp. 255–89; *idem*, *Mining in the East Midlands*, pp. 309–10.

[85] See below, pp. 214–7.

Moreover, as in the case of Cable Street, the significance of the victory at Harworth may be mythical as well as legendary. The NMFU was a genuine compromise between the NMA and Spencerism; and George Spencer was more than an honorary President of the new Union. The Managing Director of the Bolsover Company reported in 1938 that 'Mr. Spencer continues in office as President and the Officials recruited from the two old Unions are working harmoniously under him for the good of the members and the Industry in general.'[86] Even in the 1940s, the weakness of unionism in the new coalfield was still apparent: the 'political' membership (i.e. those paying the Labour party levy) at Thoresby was twenty-five out of 1,385 employed, at Clipstone thirty-two out of 1,207, at Ollerton and Harworth well under 50 per cent even of union members.[87]

Thus we may see that the economic and social effects of the opening of new collieries cannot be divorced from the political consequences. The isolation and social dislocation of the company villages of the new Dukeries coalfield reinforced the traditional moderation of naturally prosperous Nottinghamshire to make possible the most serious incidence of non-political trade unionism in twentieth-century Britain. Nor should we forget the hidden strength of non-unionism. The lack of conventional industrial and political organization in the new mines provides one more example of the challenge the coalfield of the Dukeries offers to the established view of mining life and community.

[86] Bolsover Company Annual Reports, 1938.
[87] NCB Notts Area Political Minute Book, 1947–.

6. Party Politics and Local Government

'The last of the laggards . . .'[1]

The Spencer union, dominant in the new pits of the Dukeries coalfield in the 1920s and 1930s, was hardly an ideal vehicle for organizing the miners and expressing dissatisfaction with their lot in pit and company village alike. This arm of the labour movement in the new field scarcely fits in with an interpretation of 'traditional proletarian' miners adopting a strong class view of society.[2] Spencerism was backward looking, bureaucratic, and believed in co-operation with the employers rather than in adopting 'political' trade unionism and socialist policies. What of the other arm, the Labour party, of which miners and mining areas are usually considered such solid supporters? After all, many of the parliamentary seats the young party had won by the early 1930s were situated on the coalfields, seats which remained loyal even in 1931 and which still gave Labour landslide majorities at general elections. Many municipal councils in British coal-mining districts had fallen under the control of Labour councillors. Once again, though, the Nottinghamshire coalfield shows a startling break with convention in the inter-war years.

It must first of all be stressed that Labour was substantially represented on the Nottinghamshire County Council throughout the inter-war period, with over twenty councillors by 1928, and twenty-seven in 1937—almost exactly a third of the council, including aldermen.[3] The old coalfield on the border with Derbyshire had been electing Labour councillors since before the First World War, and established mining

[1] See below, p. 132.

[2] David Lockwood, 'Sources of variation in working class images of society', *Sociological Review*, NS 14 (1966), 249-67; Martin Bulmer, 'Sociological Models of the Mining Community', *Sociological Review*, NS 23 (1975), 61, 92.

[3] B. F. L. Long, 'A study of the membership of selected local authorities with specific reference to social and political change', Nottingham University MA thesis 1964, pp. 183, 196.

towns like Mansfield and Worksop consistently elected Labour-controlled councils in the 1920s and 1930s. The old Nottinghamshire coalfield may have been what Roy Gregory called a 'laggard' in its change of allegiance from the Liberal party to Labour,[4] and it may have been renowned for its moderation and its lack of militancy, but functioning Labour party branches were operating after the First World War everywhere in the old coalfield, and the Labour group formed a recognizable opposition to the controlling anti-Socialist group on the County Council. Yet the new Sherwood Forest field in the east of the county elected but a single Labour county councillor before 1946,[5] and in the Southwell Rural District, the local authority in which most of the new colliery villages were situated, there was no Labour representative on the council until the 1946 elections.

If Gregory's classification of the various British coalfields as 'front runners', 'slow starters', and 'laggards' in coming to support the Labour party is accepted, the Dukeries must be seen as very much the last of the laggards. No Labour party branch was founded in the Sherwood Forest mining villages until the 1940s, twenty years after the economic transformation of the area brought about by the sinking of the pits. Even at Harworth, which in many ways was more typical of the neighbouring South Yorkshire coalfield than that of Nottinghamshire, and which was to lead the fight against the Spencer union in the 1930s, the Labour party hardly got off the ground in the early years of the village. At the AGM of the Bassetlaw Divisional Labour Party in 1934 it was revealed that the Bircotes branch had contributed no individual membership subscriptions for the previous year,[6] and in 1933 the constituency Executive Committee had made a special appeal for people to come from Bircotes to a meeting organized by the party at which the Labour parliamentary leader George Lansbury was to speak.[7]

[4] Roy Gregory, *The Miners and British Politics* (Oxford, 1968), p. 144.

[5] At Clipstone in 1937—see below, p. 158. Harworth, which was situated in the north of the county in the Worksop Rural District, is not considered in this paragraph.

[6] Bassetlaw Constituency Labour party Minute Book, AGM 20 Jan. 1934.

[7] Ibid., Executive Committee 8 Apr. 1933.

During the 1920s and 1930s the Conservative or Independent candidates for local authority seats were usually returned unopposed. At the lowest level of local government, the parish councils of the villages at which pits were sunk remained under the control of the colliery management and the old local leaders throughout the inter-war period. Usually the parish councils were not elected by secret ballot, but at open parish meetings by a show of hands. Naturally this inhibited opposition on the part of the colliery employees when members of the management were present at the meetings, often themselves standing as candidates. Often the parish meetings were ill attended—at Ollerton in 1934 only sixteen electors were present[8]—and it was far from unknown for parish councils to be elected unopposed, even when there were as many as a dozen vacancies: at Boughton in 1934 there were only nine nominations for the ten seats.[9] The miners shied away from an involvement in politics which would be unpopular with the coal companies which had so much influence in the villages.

Despite the numerical predominance of miners in the electorate (the colliery villagers always outnumbered the 'old villagers' by at least two to one),[10] even when there was a contest in the pre-war years it almost invariably resulted in a heavy Labour defeat. In 1937 the overwhelmingly mining ward of Bilsthorpe elected the Conservative Lady Savile by 763 votes to the 389 polled by the Labour candidate, an ex-miner.[11] As late as 1935, over ten years after miners first arrived in the Dukeries in great numbers, the Southwell Rural District councillors for Ollerton were colliery manager Monty Wright, Chairman of the New Ollerton Conservative Association, and J. F. Jones, landlord of the Hop Pole Hotel. At Clipstone the councillors were colliery manager W. Goodman and the local Conservative branch chairman, the solicitor Frank Armstrong. At Bilsthorpe the representative was its colliery manager, L. T. Linley; at Edwinstowe, colliery under-manager W. Tagg and garage proprietor Percy Morley.

As Table 27 shows,[12] the representation of the new

[8] *Dukeries Advertiser*, 6 Apr. 1934.

[9] Ibid.

[10] See below, Table 32.

[11] *Nottingham Guardian*, 6 Mar. 1937.

[12] See below, p. 157.

mining villages tended to be shared before the Second World War by an official of the local coal company, usually the colliery manager, and a leading member of the old village, often someone with close connections with the Dukeries aristocracy, who still owned most of the land in the district. Percy Morley of Edwinstowe, for example, a prominent figure in the 'old village', had many dealings with the great landowners, as one of his employees reported:

RJW What was Percy Morley like?
GC I worked for Percy Morley for quite a number of years. He didn't pay very well. That was one of his traits, passed on to his son. He dealt with the aristocracy more than anyone else. He ran a fleet of Daimler cars. Five Daimlers.
RJW From Edwinstowe?
GC Yes, in the High Street . . . they were for hire, you see, and they were not only for hire locally for the aristocracy. See the time when Lady Manvers wanted taking into Retford and she had a Daimler of her own, and it was wet, she wouldn't have it out, she'd hire one of Morley's. I've done that job. And being the centre of the Dukeries, we were involved at Morley's with the aristocracy as such, probably more so than we were with the colliery village because it wasn't, well, blessed with motor cars, was it?[13]

The electoral co-operation between the colliery owners and the leaders of the 'old villages' was reminiscent of the consultation between the companies and the estate officials over other matters of mutual interest. There was certainly something of a closed circle as far as the representation of the villages was concerned. When Monty Wright resigned as the Ollerton representative on the Nottinghamshire County Council in May 1938, the local newspaper reported that he had 'asked Major John Hole to take his place. He was sure that Major Hole was "the right sort of chap" to take on the work.'[14] As Major Hole did not accept the nomination, Wright was succeeded by F. E. Jones of the Hop Pole Hotel, Ollerton.[15]

Whether or not they stood in the local council elections as 'Independents', there was often a direct connection between

[13] Interview, George Cocker, Edwinstowe.
[14] *Ollerton, Edwinstowe and Bilsthorpe Times*, 6 May 1938.
[15] Ibid., 20 May 1938.

the colliery management and the Conservative party. As has already been noted, it was not unknown for the colliery manager to be the chairman of the Conservative party branch in the Dukeries villages. When L. T. Linley, first manager of Bilsthorpe Colliery (1925-36), died on 29 April 1941, the *Worksop Guardian* described him: Vicar's Warden of the new parish of St. Luke's, a school manager, a keen worker for the St. John's Ambulance Brigade, and 'a lifelong Conservative, he was always active and anxious to help the cause.'[16]

The economic and social dominance of the coal companies and the local landowners undoubtedly played a major role in the lack of Labour party activity in the Dukeries before the Second World War, for opposition was a serious business:

RJW Was there a Labour party organized in Bilsthorpe before the war?

HT Now then, when we first came here in 1927, the County Councillor for this area—but we start with the MP. The MP for this area were the Marquess of Titchfield, with the renowned name 'the silent MP'. According to records the only thing that he ever mentioned in Parliament was 'close that window' where a draught is, that's the old folklore. The County Councillor was Earl Manvers, big landowner. The District Councillors was management, kind of thing—

RJW Linley was District Councillor.

HT —Linley if you like, management. The Parish Council was all management. You see nobody dare oppose people, nobody dare oppose. And it was rank Conservative right up to 1944.

RJW So, although this was a mining village, and almost the whole population were miners, it was Conservatives unopposed, for twenty years after the—

HT Nobody dare put up! That's right. No doubt about it. They were opposed but in a very very unorganized way, before that. Because as I say there was this fear of opposing in every colliery village around here. It wasn't just Bilsthorpe, it was all round. Now then, in 1944, when the war was on, then we started battling for freedom if you like, one or two of us in every village all round here, mining villages, and became organized. Funnily enough there was no teamwork. It was done with half a dozen men in each village without the other half dozen knowing, but it was done in around 1943-4.[17]

The Conservative party domination applied also at the level of parliamentary representation. The new coalfield was

[16] *Ollerton, Edwinstowe and Bilsthorpe Echo*, 2 May 1941.

[17] Interview, H. Tuck, Bilsthorpe.

mainly in the constituency of Newark, whose member from 1922 to 1943 was the Marquess of Titchfield, son of the Duke and Duchess of Portland of Welbeck Abbey.[18] The Nottinghamshire mining seats of the old coalfield, Broxtowe and Mansfield, remained loyal to Labour even in the disastrous year of 1931, but at Newark the Labour vote grew only slowly in the inter-war years. In 1922, before miners had arrived in the Dukeries in any numbers, the Labour vote at Newark stood at just over 8,000. Most of these would come from the semi-industrial town of Newark-on-Trent, where there was a functioning Labour party organization. In 1929, the year of the last general election before the 1930s with its National Governments, the Labour vote still stood at 8,000, but now there were nearly 11,000 Liberal voters, and the Marquess of Titchfield was elected with a minority vote of 15,707. It seems clear that many of those who did not vote for Titchfield in 1929 were miners from the Dukeries colliery villages, although we have no way of telling how the different parts of the constituency voted in the general election. It should not be surprising that miners should vote Labour in general elections when given the chance, even if, as is possible, many of the immigrants to the Dukeries in the 1920s were voting for the first time (the turnout in 1918 had been very low). There is a great difference betweenn on the one hand, voting Labour in a constituency which spread beyond the boundaries of the colliery village, and, on the other hand, being seen to organize a local Labour party or even to stand against the colliery manager and the old élite in a municipal contest.

The general elections of the 1920s and 1930s are of less use than the local elections in trying to explain the peculiar nature of party politics in the Dukeries villages for three reasons: because it is hard to interpret the results of the diverse seat of Newark which contained agricultural, urban, and mining areas; because there is little hint in the general election figures of the lack of Labour party activity in the colliery villages and of the reason for it; and because the new coalfield's development did not in any case lead to a very

[18] Harworth and Langold were in Bassetlaw constituency; Blidworth was in Mansfield constituency.

significant increase in the overall electorate in the Newark constituency. Between 1926 and 1931 the colliery villages accounted for about 4,500 new electors in this division, but even this modest impact must have been reduced by the way the electoral roll would have been inaccurate, due to the rapid turnover of population. The increase in the electorate between 1924 and 1929 was mainly due to the extension of the franchise to women under 30 in 1928, and is paralleled in other British constituencies.[19] All in all, it is hard to draw any firm conclusions from the parliamentary election figures of the period. There may have been a lower turnout in general elections in Ollerton, Edwinstowe, Clipstone, and Bilsthorpe than in other British mining villages or than other parts of the constituency. There may have been proportionately a lower Labour vote than in long-established mining communities. But this cannot be shown by general election results. It is clear, on the other hand, that miners drew back from forming local Labour party branches in the colliery villages, that they drew back from fighting parish, district, and county council seats, and that in the few contested municipal elections there was a reluctance to vote Labour before the Second World War.

After 1945 the local election picture changed dramatically, and after 1945 too the parliamentary seat of Newark fell to Labour for the first time. Even though it had been held for the Conservatives in Labour's landslide year of 1945, Newark was taken easily in 1950 by George Deer. The boundaries had been changed to include the long-established Labour stronghold of Mansfield Woodhouse; in 1955 the Boundary Commision returned Mansfield Woodhouse to Mansfield constituency, yet Labour held Newark, as they did in every subsequent general election until 1979. The only difference between the boundaries of the parliamentary seat in the 1950s and the 1920s was the exclusion of the small Bingham Rural District, certainly a Conservative area, but unlikely to account on its own for the change in the constituency's

[19] The national increase in the electorate between 1924 and 1929 was 32.8 per cent. In Newark constituency the increase was 42.5 per cent, from 31,458 to 44,826. Of this 4,326 was accounted for by parishes undergoing colliery development.

allegiance from Conservative to Labour. Probably most of the Dukeries miners did vote Labour in the post-war general elections, helping Labour to win the seat. However the general election results are as unreliable a guide to the politics of the new coalfield in the years after the war as in those before, and the boundary changes make it even more difficult to draw conclusions. Other types of evidence illustrate more decisively the lack of Labour party activity in the Dukeries colliery villages before the war, and its rapid growth in the 1940s.

As far as the Liberal party in the Newark constituency was concerned, it was clearly capable of making a good showing at the 1923 general election, when the Liberal candidate lost to the Marquess of Titchfield by 2,616 votes in a straight fight, and in 1929, after the arrival of mining in the Dukeries, when J. Haslam finished second in a three-cornered contest. There was a functioning and effective Liberal organisation in the non-mining section of the constituency, in Newark-on-Trent borough and in the agricultural villages. However it is impossible to discern any significant impact on the Liberal vote resulting from the influx of miners, which, as has been pointed out, was too small to allow valid interpretation of parliamentary elections. It can only be said that no Liberal candidate stood in local elections in the new coalfield before the Second World War, nor indeed for many years after it. Nor were Liberal branches founded in any of the new colliery villages in the 1920s and 1930s, and if any Liberals migrated from older coalfields to the Dukeries they have left no trace of political activity in their new homes.

How do we explain the lack of Labour party activity in the Dukeries mining villages before the Second World War? First we must consider the lack of party organization in the new coalfield. No ward or branch parties were founded in the new colliery villages, and activity was confined to the occasional informal effort by a group of like-minded friends.[20] Even the deeply conservative market town of Newark-on-Trent itself had an operating Labour ward organization in the inter-war period, with half a dozen town councillors in a good

[20] Such as Bilsthorpe in 1937. See above, p. 135, and below, p. 158.

year,[21] whereas the activity in the mining part of the constituency was negligible. To some extent this was due to the fear of alienating the colliery management, which prevented most men from helping to organize a Labour party in the new villages. The hostility of the colliery management to socialism was clear. Not only did colliery officials take a leading part in local government and local Conservative associations, but coal-owners did not draw back from speaking to the miners in an overtly political manner. When Butterley Company Director Captain Fitzherbert Wright opened the temporary Miners' Institute at Ollerton in July 1924, he took the opportunity to comment on the first Labour government:

> There was a movement now to put an end to private enterprise and private ownership. The idea was abroad that those responsible for private enterprise (and those undertaking big enterprises like the Ollerton Colliery) got too much out of it. Some people also thought that because directors and managers rode in a Rolls Royce car—as he did—that they paid small wages and so on, but he was of the same opinion as Mr. Wheatley who in the House of Commons the previous day said England was not prepared to adopt Socialism at the present time. (Applause). The people still believed in private enterprise, individual liberty, and individual ownership of property. (Applause). At any rate, the Company intended to continue with their efforts for the development of the coal industry in Notts and Derbyshire until turned out. (Applause).[22]

The poor Labour organization in the Dukeries contrasted sharply with that of their opponents. The Conservative party were fully aware of the threats posed to their traditional dominance in the area by the changing character of the electorate, and tried many methods of combating the danger. In 1927, for example, the Junior Carlton Club's travelling cinema van visited Clipstone and Ollerton.[23] Men's and women's party branches were founded in the new villages, usually led by coal company officials and their wives in alliance with the leaders of traditional Dukeries society. The annual Conservative constituency party meeting in June 1927 reported that:

> They had paid particular attention to the new colliery districts of Ollerton

[21] *Newark Advertiser*, *passim*.
[22] *Worksop Guardian*, 1 Aug. 1924.
[23] *Newark Advertiser*, 20 June 1927.

and Clipstone, where there had been an influx of new voters. They had held a series of successful smoking concerts there each succeeding one being better attended than the previous one. Lord Titchfield had met with a cordial response at these places. (Applause). There were now fully equipped men's and women's associations at Ollerton, and the work in that district had gone very well.[24]

In January 1928 the Marquess was again given a 'cordial welcome' at a Clipstone smoking concert in the colliery village's recreation hut. The object of his visit was to form a local branch of the Newark Conservative Association, the president of which was Clipstone colliery manager W. H. Mein and the vice-president the solicitor F. Armstrong, who represented Clipstone on the Southwell RDC.[25] In August 1929 we hear that the Conservatives were 'still working hard in the Colliery Districts, and during the year a Men's Association had been formed at Bilsthorpe.'[26] The activities of other branches such as those at New Ollerton and Edwinstowe were also extensively reported by the virulently pro-Conservative local press. There can be no doubt that the Conservative party responded to the new political situation in the Dukeries with vigour and efficiency. It must be added, though, that in many ways their task was made easier by other forces tending to lend support to Conservatism in the new coalfield.

The political temperament of the district cannot be understood without taking into account that the coalfield was situated in the heart of one of the most traditionally-minded, rural regions of the country, justifiably known as the 'Dukeries'. The new pits were sunk on the estates of the Dukes of Portland and Newcastle, Earl Manvers, and Lord Savile, who owned houses at Welbeck, Clumber, Thoresby, and Rufford, respectively. Between them these magnates owned vast tracts of land in the 1920s. Traditionally this part of Nottinghamshire was a stronghold of deferential voting and aristocratic pocket boroughs. Earl Manvers had been Conservative MP for Newark as Lord Newark from 1885 to 1900. W. E. Gladstone had first entered Parliament in 1832 as the member for Newark owing to his friendship with Lord Lincoln, son of the Duke of Newcastle, who controlled

[24] Ibid., 20 June 1927. [25] Ibid., 25 Jan. 1928. [26] Ibid., 7 Aug. 1929.

the seat. The respect even amongst the miners for these families cannot be ignored, and the Marquess of Titchfield certainly benefited from the widespread reverence for his mother, the Duchess of Portland, who was famed for her 'outspoken and impassioned work on behalf of overburdened persons, horses, the maimed and sick and animals of all sorts.'[27]

Deference, however, cannot be clearly distinguished from the economic power the local landowners enjoyed. Two interviewees from Edwinstowe described how the Marquess of Titchfield was regarded:

RJW Now what did people think of him in Edwinstowe? Did they respect him? Did they vote for him?
FT They didn't know any better! I should think if ten per cent of the people in the village knew him when they saw him, that was about it. They'd probably never seen him.
GC That was touching the forelock days, that was.
RJW What did people think of the aristocracy in general?
GC Their bread and butter came from them, so they had to respect them.
FT They'd no choice.
RJW In what sense do you mean their bread and butter came from them? Do you mean the estates employed them?
FT The Thoresby estate, the Welbeck estate.
GC They owned the village.
FT Places and properties. It was, I don't know, a type of dictatorship, it had got to be, hadn't it?[28]

This view was corroborated by an interviewee from the old village of Ollerton who had previously lived in the Thoresby estate village of Budby:

RJW Would you say that most of the people in the old village of Ollerton here would support the Marquess of Titchfield as their local MP?
AN Support him as an MP? Well, you see there again, if you wasn't a Conservative on these estates, if you were Labour you had to say you was Conservative. It wouldn't do you any good. If you were going to vote for a brainy person you wouldn't put him as bright!
RJW Who, Titchfield?
AN Yes. The only thing, if you were a Conservative, he was working for us, but you wouldn't vote him for his values.

[27] Interview, A. E. Corke, Ollerton.
[28] Interview, F. Tyndall, G. Cocker, Edwinstowe.

RJW Do you reckon there was any pressure from John Baker here and Lord Savile?
AN Oh, for goodness sake, yes.
RJW In the village, and not just on the estate?
AN Oh yes, because a lot of the village people worked for them, or Thoresby. They were all Conservatives.[29]

When assessing the landowners' political influence in the Dukeries, we must bear in mind that not only did they employ the bulk of the residents of the old villages on their estates, but we should consider the view that men living on estates simply did not believe that the ballot was effectively secret even in the 1920s and 1930s.[30]

The inhabitants of the district who had known its fine countryside before the arrival of pit chimneys and spoil heaps were united behind the Conservative party by fear, not so much of potential damage to the environment as of the social and political perils occasioned by the coming of the miners. There was little friendly contact between the 'old villagers' and the residents of the colliery villages, and in order to try to maintain their existing ways of life in the non-miners united more firmly than ever behind the Conservative party, still led by the old landed and professional classes, although these were now in alliance with the coal company management.

There is some evidence that men from less traditionally moderate parts of the country brought additional political militancy to the new Nottinghamshire coalfield. The campaign against the Spencer union at Harworth in the mid-1930s was widely regarded as the work of itinerant agitators like the Scot, Mick Kane. Kane had been elected to Worksop Rural District Council in 1936 and was actually a serving councillor at the time of his trial for intimidation of strike breakers in 1937.[31] (He was found guilty and sentenced to two years hard labour.) Harworth was not in the Dukeries proper but on the Notts.-Yorkshire border, and it was certainly the most heterogeneous of the new pits, which perhaps accounts for its divergence from the pattern of labour inactivity on the

[29] Interview, Albert Nuttall, Ollerton.

[30] P. R. Shorter, 'Electoral Politics and Political Change in the East Midlands of England', Cambridge Ph.D. 1975, p. 49.

[31] *Worksop Guardian*, 16 Apr. 1937.

new coalfield. The Todhunter family, who provided a Labour and NMA candidate against Lady Savile at Bilsthorpe in 1937, hailed from the North East.[32]

The founder member of the Labour party in Bilsthorpe was one of the Durham miners who had come down here, and he'd got about six sons . . . Todhunter was the man, but we rallied round him, and there was myself and one or two of Todhunter's sons, and I would say that there were no more than half a dozen of us formed the party.[33]

The former Clerk to the Parish Council at Ollerton, J. E. Smith, held that the beginnings of Labour party activity in the village were brought about by a group of Yorkshire miners during the war, when they demanded that the Parish Council should be elected by poll instead of by show of hands at the annual parish meeting:

If six parishioners get together, and they attend the annual parish meeting, and they demand a poll, by virtue of the law a poll must be held. And it was around '42 or '43 when the annual parish meeting this was mooted, and ever since then it has been by poll; prior to that, it was always by show of hands.[34]

This fits in with evidence from other fields: Leo Loubère suggested that migrant newcomers led a political movement against company control in Lower Languedoc in the late nineteenth century:

The presence of a large non-native population had a marked influence on group behaviour. For example, the town of Bessèges, with nearly 50% of its inhabitants from other departments, changed from a centre of royalism in 1848 to one of revolutionary syndicalism in 1890. Grand'Combe, with nearly two-thirds of its inhabitants native-born, remained strongly royalist until the 20th century.[35]

Miners immigrating from the Hérault and Gard could be seen taking the lead in industrial negotiations, just as long-distance immigrants could in the Dukeries. Long-distance immigrants working in the burgeoning car industry in Oxford between the world wars seem to have taken a lead in militant union activity, clearly partly because they originated from a background

[32] Lord Taylor, *Uphill all the Way* (London, 1972), p. 60.
[33] Interview, H. Tuck, Bilsthorpe. [34] Interview, J. E. Smith, Ollerton.
[35] Leo Loubère, 'Coal Miners, Strikes and Politics in the Lower Languedoc 1880–1914', *Journal of Social History*, 2 (1968), 27.

more used to trade unionism than those who came from surrounding agricultural villages or than native Oxonians employed in service or distributive trades.[36] In general, however, the immigrants, whatever their origin, bowed before the prevailing circumstances in the coalfield—political quiescence, paternalism, and a difficulty of organization born of company power.

We might also consider the connection between the Spencer union and the Conservative party as a possible handicap to Labour party political activity. No formal financial link may be traced between the NMIU and the Conservatives, but friendly relations at an individual level were common: in January 1928, at the annual dinner of the Ollerton Colliery Sports Club, the Marquess of Titchfield signified his intention of becoming a trades unionist. According to the *Newark Advertiser*, he said, 'I am going to ask you a great favour. I am going to ask to be allowed to join your Union—after all, I am your representative in the affairs of State, and I get my living from the coal trade as you do.'[37] The *Advertiser* did not see the need to mention at any point that the union concerned was the NMIU, the Spencer union. Again in 1928, at a smoking concert in Clipstone, the Marquess said he was a strong supporter of the new trades union that the miners had formed because it had recognized those principles which were the true principles of the Conservative party, 'personal possessions, religion and the right of men to work out their own salvation in the right way . . . miners could translate those principles into action through the work that their great leader, George Spencer had started.'[38] On occasion the Conservative party's advances to the Spencer union were reciprocated; Joseph Shooter, who was to be first Branch Chairman of the NMIU at Rufford, proposed a vote of thanks on Baldwin's visit to the Portland seat at Welbeck with the words 'On, Stanley, on with your splendid endeavours to serve our King, Country and Empire.'[39] The union secretary at Clipstone is said to have had a big poster in his window at election time—'Vote Conservative. Vote for the Marquess of

[36] See below, ch. 12.
[37] *Newark Advertiser*, 11 Jan. 1928.
[38] Ibid., 25 Jan. 1928.
[39] Ibid., 8 Jan. 1925.

Titchfield.'[40] Spencer was of course himself a Labour MP (for Broxtowe, now Ashfield) until 1927, when he resigned the whip. Generally speaking, the Spencer union's view that men and management were natural allies, their co-operation with the employers, and their dominance in the new coalfield could not have made the task of forming a Labour party in the Dukeries any easier. For all these reasons the turnout of the miners in municipal elections was low and their organization poor. As a result, the Conservative-Independents managed to avoid ceding control to the Labour party in the new coalfield before the social and political upheaval of the Second World War.

'Trying to mix oil and water': Local government in the new Dukeries coalfield

Why did the colliery companies become interested and involved in local government in the Dukeries in the 1920s and 1930s? Why did they frown upon Labour party activity amongst their workmen? It is essential first to consider the problems that the local authorities faced in coping with the arrival of coal-mining in their districts. Harworth Colliery (Bircotes village) and Firbeck Colliery (Langold village) were situated within the Worksop Rural District, which was known until 1925 as the Blyth and Cuckney Rural District. Of the other new pits sunk in Nottinghamshire in the 1920s, Clipstone, Bilsthorpe, Thoresby-Edwinstowe, and Ollerton were always in the Southwell Rural District. Blidworth was part of the Skegby Rural District until 1933, when the Skegby RD was abolished and Blidworth parish transferred to Southwell.

Both Worksop and Southwell RDs were previously completely agricultural in nature, and councillors from grossly over-represented agricultural parishes continued to outnumber those from the parishes growing rapidly as a result of colliery departments. The councils were entirely unaccustomed to coping with the problems caused by the transformation of the neighbourhood due to the arrival of mining and many thousands of miners. As the part-time Clerk to Southwell

[40] L. Dowen, Clipstone, interview with P. A. Turner.

Rural District Council put it, when the Ministry of Health sanctioned a grant of £40,500 for the development scheme involving 548 colliery houses at Edwinstowe in 1924, 'this was the biggest thing the council had ever undertaken'.[41] Although almost all the houses for the new population were built by the colliery companies and not by the local authority, the councils did become involved owing to the administration of the government housing subsidy scheme. Furthermore, they were directly responsible for providing water and sewerage schemes for the new villages, and for local roads. This involved the councils in expensive undertakings for the first time—for example, in 1926, the Southwell Council had to sanction a housing subsidy loan of £10,500, £23,000 for a sewage disposal scheme for Bilsthorpe, and a water scheme for the same village costing £13,850.[42] They were constantly being trapped by the fact that their borrowing capacity was related to the rateable value of the district, which only increased after the schemes for which they had to pay were completed.

The council members were in general most unsuitable candidates for the job of dealing with the problems caused by the arrival of coal-mining. Representatives of rural parishes and farmers formed the largest group of members, and often they resented the increases in rates necessary to pay for the costly schemes for providing the amenities and services required by the new mining communities. Frank Armstrong of Clipstone was a particularly consistent critic of the 'monstrous extravagance' of the Southwell Rural District Council,[43] and of the housing subsidies paid through the local authority. Even in the 1950s the representatives of the coal coal-mining villages still had to struggle with the problem of the majority of councillors from agricultural areas. As a former Southwell councillor put it,

> One had to resort to devices at times. One particularly sticky question I know we were agreed to have deferred because we knew full well there would be some haymaking doing and that our opposition wouldn't be in such a good voting strength. It worked all right, but actually we

[41] *Nottingham Journal*, 22 Nov. 1924.
[42] *Newark Advertiser*, 5 May 1926.
[43] Ibid., 17 Aug. 1930.

were in a minority, it didn't maintain at the adoption of the committee minutes by the full council—it was referred back and defeated.[44]

Personnel on the council tended to remain unaltered for many years. At Southwell, Alexander Straw was elected chairman each year from 1923 to 1945, while John Ellis, only a part-time Clerk to the Council, celebrated twenty-five years' service in 1934.[45] In the Worksop Rural District, William Ghest remained chairman until 1935, by which time at the age of 93 he was the oldest council chairman in the country.[46]

Southwell RDC in particular infuriated both the county council and the colliery companies by its slowness and inefficiency. There was a notorious row in the summer of 1927 when the already belated opening of a new school for the colliery village of New Ollerton at Whinney Lane was further delayed by the council's failure to provide water and sewerage services for the school or to metal the lane connecting the school with the village. The Chairman of Notts. County Council Education Committee, Alderman Edge, accused the Southwell RDC of negligence. As the *Nottingham Journal* of 28 June 1927 reported:

Alderman Edge said that there were about 400 children running about wild. They could not open the school without water and a drainage place for the sewage, and that was entirely due to the absolute neglect of the Southwell RDC who had not carried out the work they knew they had to do. The fact of the matter was that there were large colliery buildings going up in the district, and they were not competent for the job before them.[47]

Part of the problem was that the Southwell Council, at the instigation of the 'die-hard' anti-spender Armstrong, had tried to get away with an 18 foot carriageway for Whinney Lane rather than the 24 foot asked for, in order to cut the cost of the scheme from £8,000 to £6,000, but the Ministry of Transport had overruled them and forced them to adopt the more expensive plan against their will.[48] The Rural District Council also crossed the business-like Butterley Company on

[44] Interview, A. E. Corke, Ollerton.
[45] *Nottingham Journal*, 28 Apr. 1934.
[46] *Retford, Gainsborough and Worksop Times*, 25 Apr. 1935.
[47] *Nottingham Journal*, 28 June 1927.
[48] *Newark Advertiser*, 27 Apr. 1927; ibid., 3 Aug. 1927.

various occasions and attracted many complaints in the company files. As mining agent H. E. Mitton put it in August 1927, 'they have been slack enough to allow an overdraft to accumulate while they still have borrowing powers. I think it is high time that the whole of the councillors were relieved of their office.'[49] As a result, the Butterley Company had to lend the Southwell RDC £10,721 to cover the cost of the sewerage scheme for the new village of Ollerton. The company then tried to speed the repayment of the money but found that the Clerk to the Council was 'very dilatory, and does not carry out his engagements'.[50] Similar problems concerning widening and improving the roads in Ollerton led Mitton to describe the local authorities as 'in a fog' in January 1925,[51] but this time Butterley refused to pay for any of the cost. As late as October 1943 it was remarked that 'we know that the Southwell RDC is not carried on with what might be called the highest standard of efficiency.'[52]

Indeed, the coal companies' efforts to influence the plans and decisions of the rural district councils went beyond mere exhortation. On various occasions the Butterley Company took the lead in attempting to do something positive towards improving the efficiency of local government in the Dukeries by taking the new coal-mining villages out of the Southwell Rural District and establishing a new urban district council. The main attempt of this nature took place in 1926. Butterley mining agent Eustace Mitton initiated the idea. On 7 October 1926 he wrote to the company's lawyers, Thicknesse and Hull:

> I have not had any further communication from the Clerk to the Southwell RDC in response to my repeated applications for something to be done, therefore I shall be glad to discuss the matter . . . and the question of approaching the Government with a view to obtaining Urban Powers for this area.

It is interesting that at this stage Mitton was not contemplating bringing all the new mining villages together into an urban district council, but only the area immediately round Ollerton, that within Butterley's 'sphere of influence', where they

[49] Butterley W1/1 433.
[50] Ibid.
[51] Butterley E4/11 156.
[52] Butterley E3/4 156.

would not have to cope with the authority of a rival coal company: 'This company are so dissatisfied with progress in connection with the development of the Ollerton District, an and they are seeking powers for Urban District Authority for Ollerton, Wellow and Boughton, in order that progress can be made.'[53] On 4 November Mitton wrote to Sir Ernest Gowers of the Mines Department asking for support for his proposals, as the Rural District Council was 'hopelessly behind in giving the necessary requirement to the inhabitants in the matters of Drainage, Water etc.' When it was suggested that the proposed new urban district council was too small, Mitton added the villages of Ompton, Kirton, and Walesby to his plans; still he was reluctant to include the new mining villages in one authority. However, Butterley's complaints concerning Ollerton were reinforced by opinion in other new colliery undertakings.

T. Warner Turner, County Council representative for the Edwinstowe division, added that

> at Clipstone, where the Duke[54] owns most of the land, they had been unable to get on with the Drainage and necessary water supply etc. for the inhabitants, simply because he could not get anything at all done by the Rural Council. At Clipstone the Colliery was commenced in 1913, and today they have over 700 houses—they had no Drainage, and the land was saturated with sewage, and it was a public disgrace, and unless something was done very shortly there would be an epidemic.[55]

The coal companies' attitude to the Southwell RDC is best summed up in the Case for Counsel prepared for the Butterley Company in December 1926:

> The development of the Coalfield is creating an entirely new set of conditions in the District. Whereas in the older coalfields the development of Coal Mining meant additional housing in an already populated district, the development of the new coalfield involves the creation of entirely new towns and villages and the installation of large and expensive schemes of Sewerage, Water Supply, Lighting and street making. The Colliery Companies on their part are spending large sums in building but they are and will be hampered by the difficulties in getting the Local Authorities to perform their statutory obligations. Ollerton is in the Rural District of Southwell. This is a large district and until recently was entirely agricultural. The Rural District Council, which till lately

[53] Ibid., N5 415, Mitton to John Baker, 15 Oct. 1926.
[54] The Duke of Portland. [55] Butterley H1 415.

had only to manage the affairs of a small cathedral town and a number of rural villages has now to undertake the duty of providing local government facilities for a number of large colliery villages which are springing up and which will bring a very substantial population into the district.[56]

However, the Butterley application was not successful, and the new mining villages never did attain Urban status, although the proposals of 1926 were renewed in 1930 and 1943. One success in which the Butterley Company was involved was the redrawing of the ecclesiastical parish boundaries in the Ollerton area to place the model village in one parish, instead of being divided between Ollerton and Boughton parishes as formerly. The company argued that the social homogeneity of parishes should be maintained: the colliery housing should not be split up, while 'when it comes to matters of street lighting, sewerage, water supply etc the old inhabitants of a parish like Boughton do not like paying extra rates for the New Village.'[57] As a Boughton parish representative put it, 'they were agricultural people, and to ask them to join with the mining fraternity at Ollerton was like trying to mix oil and water.'[58]

It is interesting to note that when Southwell RDC Clerk John Ellis wrote to Monty Wright asking him if he had any suggestions concerning the review of parish boundaries, Wright replied that he would put the matter before the Butterley Board of Directors, whereupon Ellis felt obliged to remind him that he had written to Wright in his capacity as a rural district councillor for Ollerton rather than as manager of Ollerton Colliery. This exemplifies Wright's view of his role as councillor as part of his duties in protecting the coal company's interest in and control of the village. It is scarcely surprising that they felt their investment threatened by the local authority's inefficiency—a weak link in the chain of their exploitation of the new coal resources of the Dukeries. Nevertheless, inefficiency was not likely to be as dangerous to the coal companies as Labour party activity in the Dukeries, and the companies did use their considerable influence in the colliery villages to secure the re-election of Conservative and

[56] Ibid., H1 415.

[57] L5/2 231, Wright to Ellis, Nov. 1929.

[58] Ibid.

Independent councillors—a policy pursued with general success until the 1940s.

'This is the end of the Dukeries, the passing of an age'[59]

After the war, the Labour party won every election they contested in the Dukeries mining villages, until the 1970s, when increasing affluence and the greater volatility of voters led to larger swings in municipal elections. In the first local elections to be held after the war, in 1946, Labour swept the board at parish, rural district, and county council levels. In most of the 'new' mining villages—and it should be remembered that it was now about twenty years since the pits were sunk—the former parish councillors were rejected *en bloc* in favour of a completely new council made up entirely of Labour nominees.[60] There had been miners as representatives on local councils before 1946, but these had almost invariably been men acceptable to the companies and to the Conservatives, men like Spencer-union checkweighmen Aaron Jenkins (Boughton), Gil Gozzard (Edwinstowe), and Harry Willett (Edwinstowe). Willett, for example, had been a founder member of the Clipstone Conservative Association in 1928,[61] while Jenkins had replaced the sacked NMA activist Sam Booth in the controversial election in 1930 for the post of checkweighman at Ollerton Colliery.[62] However, Jenkins was also ousted from the local authority council in 1946, when the first full slates of Labour candidates contested the Dukeries mining villages. The Labour party remained in a minority on Southwell Rural District Council owing to the continued over-representation of the rural parishes, but the miners' votes helped to swing the Newark parliamentary constituency for Labour in every parliamentary election from 1950 onwards, and the Dukeries mining wards helped Labour to take control of the Nottinghamshire County Council for twenty years after 1946. Why did this dramatic change take place in the political representation of the coalfield in the 1940s?

[59] See below, p. 154. [60] See below, Table 31.
[61] *Newark Advertiser*, 8 Feb. 1928.
[62] Butterley T1/6 345, 'Notts Industrial Union 1926–45'.

Clearly the primary cause lay in the foundation of active and organized Labour parties in the new villages, which could mobilize the party's potential support in elections more effectively, and campaign for new recruits. But why did the Labour party only get off the ground twenty years after the arrival of miners in the Dukeries? To some extent the answer must lie in the destruction of those authorities which had held sway in the villages until the war years. The companies' powerful influence over the lives of their employees was broken by the security and power offered to the miners by their role in the war effort between 1939 and 1945 and their status as reserved workers.[63] There is also a well-known theory that the experience of the war years generally acted as a powerful influence for radicalization in Britain. H. Tuck of Bilsthorpe explained the current feeling that a sea-change was in the offing:

> Oh, I think definitely it was the war, yes, I have always believed this. I think the war taught the working class oppression, I think that's the word. Because we'd been oppressed, not in war, we'd been oppressed right up to that time with the private coal-owners, we'd been oppressed and whether you're a good worker or not. You weren't a militant, you didn't know what the word 'militant' is today then, although maybe we were fighters. But the point at issue was that there was a determination at the latter end of the war. When the war was finished we wasn't going back to what we had previous, I think that was the attitude.[64]

However, given that the impact of the war on the Dukeries mining villages was in general limited, and few miners gained experience of new surroundings, new comrades, and new relationships in the armed forces, it would be difficult to interpret 'radicalization' as due to anything more than an improvement in the miners' 'market position', resulting from the reduction of their fear of dismissal, combined with the destruction of the other authorities which had weakened the Labour party in the Dukeries before the war. Probably most of the miners in the colliery villages would have voted Labour in the 1920s and 1930s if given the opportunity, as in general elections. Now the obstacles placed in the path of more full and open party political involvement were removed.

[63] See below, pp. 214–7.
[64] Interview, H. Tuck, Bilsthorpe.

Nationalization, symbolized by the departure of Butterley Company Managing Director Monty Wright from his residence at Ollerton Hall, from which he had kept a paternal eye on the village since the 1920s, also led to a significant change in the political atmosphere of these 'company villages'. One interviewee from Edwinstowe illustrated it in this way:

RJW But there was no political activity in the village?
FT Very very little. The old part were Conservatives, and in the early stages of the development of the village[65] nobody knew what they was. They voted the same way as the manager if they could find out how he was going to vote, put it that way.
RJW When did that change, not till nationalization?
FT That's right, yes.
GC I think nationalization.
FT Vesting Day, that threw everything out of the window as far as Bolsover were concerned, I'll never forget boss who said to me, something or other had gone wrong, I can't remember, he says 'I suppose you know we're about to be nationalized tomorrow', you know how he used to talk,[66] and I just says 'I think so, sir', and I walked on. And you called them sir, oh yes.[67]

No longer if a miner stepped out of line would he be forced to leave the pit and hence the village. In addition to the diminution of company authority brought about by the war, and nationalization, one or two other factors should be considered which promoted the foundation of the Labour party in the Dukeries colliery villages.

The Spencer union had amalgamated with the official union, the NMA, in 1937, but George Spencer had remained president of the new Notts. Miners Federated Union. The co-operative policies of Spencerism were still very evident until 1944, when the Spencer keynote of local autonomy was weakened by the foundation of the National Union of Mine-workers with its unequivocal commitment to the Labour party. Also at about this time, the Dukeries aristocrats sold their seats and left the area, as death duties and other forms of taxation caught up with them after the nationalization of coal royalties in 1938. The great landed estates were sold or broken up. There was undoubtedly a feeling that the edifice of aristocratic leadership in the Dukeries was crumbling during

[65] The colliery village.
[66] This in a very mournful manner.
[67] Interview, F. Tyndall, G. Cocker, Edwinstowe.

the 1940s. When Sir Albert Ball decided to resell the Rufford estate on the open market in May 1938, at the first annual dinner of the joint parish councils of Ollerton and Boughton, M. F. M. Wright deplored the recent change in the ownership of the Rufford estate; this had resulted in the breaking up of the bloc which had dominated the area round Ollerton, and with which Butterley had co-operated so effectively.[68] In January 1941, the *Sunday Express* lamented that 'The Great Days of the Dukeries are Over'. The Duke of Portland had sold 10,000 acres of the Welbeck estate:

This is the end of the Dukeries, the passing of an age . . . no reason is given for the sale, but a year and a half ago, when the Duke sold 4,700 acres of his Northumberland estates and land in Ayrshire, it was admitted that the sale was due to the transfer of mining royalties to the nation under the Coal Mines Act.

The *Express* drew a grim conclusion: there had been a time when four dukes had inhabited this corner of England: Portland, Newcastle of Clumber, Norfolk at Worksop Manor, and Kingston at Thoresby Park. But now there was no longer a duke in Worksop Manor, the duchy of Kingston was extinct, Clumber demolished, and the Rufford estate sold and broken up. Did the disintegration of Welbeck imply 'the final destruction of the old order'?[69]

There was only one remaining source of opposition to Labour control of Dukeries local government. Even after Labour members were first elected to the rural district councils in 1946, there was an element of hostility shown by the farmers and traders who represented the agricultural parishes. A pioneer of Labour activity in one of the new mining villages, Bilsthorpe, reported that the first Labour representatives faced such hostility:

RJW Were they welcomed onto the Council or were they resented by the Conservatives and Independents who had dominated the Council before?

HT I think the mining community is still resented—their representatives are still resented in the Newark District, even to this day I think they are. I'm the Newark District Councillor for this village, but one feels even to this day, that we're still 'them' kind of thing,

[68] *Ollerton, Edwinstowe and Bilsthorpe Times*, 6 May 1938.
[69] Ibid., 24 Jan. 1941.

> that we are foreigners in a way of speaking, in the Newark District. Well, give you a typical example in regards to this. Under the new parliamentary boundaries, they are trying to—the Boundary Commission are trying to get a new division in this area, called the Sherwood Division, which would take all the miners away from Newark. There would be a new MP. The new Sherwood Division would stretch from Ollerton right the way down to Hucknall, right down the coal valley, if you like to regard it. Good God, it was the best thing they'd ever heard in Newark! Get shot of them! Even when we're talking about coal, even though we've six pits in the Newark area, and we're the biggest ratepayers, the Coal Board, etc. you'd never believe it, not in a Tory council. You would never believe that. You'd think the old farmer and all that was paying the money to improve the district. No, there is definitely a feeling—they may not admit to this—there is definitely the feeling in Council, that they just don't want us.[70]

It is clear that in the Dukeries it took a considerable time for the Labour party to develop, just as it took time for other institutions usually held to be typical of mining villages to appear. It is more important to point out the tardiness in founding active branches of the Labour party in the colliery villages and in putting up miners as candidates in municipal elections than to sift through the inconclusive evidence as to whether the Dukeries miners voted Labour in general elections. As has been stressed, it required considerably less commitment to *vote* for an outside Labour candidate in a secret ballot than to be seen to take the lead in opposing the management (who had good reason to care more about control of local government than paraliamentary elections). What is more, Labour voting does not reveal everything about the miners' political views. To vote Labour in Britain has not necessarily implied a commitment to socialism or to 'a criticism of the whole system of economic and social relations of capitalism'.[71] As R. S. Moore points out, the connection between Labour voting and the full adoption of 'traditional proletarian' views is by no means clear, and conversely a unitary image of society does not always involve support for the Conservative party.[72] Several of the Dukeries miners who

[70] Interview, H. Tuck, Bilsthorpe.

[71] R. S. Moore, *Pitmen, Preachers and Politics* (Cambridge, 1974), p. 19.

[72] E. Batstone, 'Deference and the ethos of small-town capitalism' in M. I. Bulmer (ed.), *Working Class Images of Society* (London, 1975), p. 118.

expressed regret at the passing of the paternalism of the private coal-owners in the Dukeries villages proved to be long-term Labour voters.

It is worth stressing that absence of Labour party activity in the colliery villages of the Dukeries could be due to deference, a positive commitment to a hierarchic view of society in which the political supremacy of the 'quality' of the village might be acknowledged as good in itself, that is 'the subscription to a moral order which endorses the individual's own political, material and social subordination'.[73] On the other hand quiescence could be caused by powerlessness in the face of the overwhelming power of the companies and the landowners.[74] It is important to notice that both elements may be found amongst the inter-war Dukeries miners. Deferential views are closely connected with the concept of the order and orderliness of the 'model' village. Powerlessness is illustrated by the inability of the miners even to put up their own candidates in municipal elections before 1946. Many miners do still show signs of resentment concerning the days of company control, and personal resentment against unpopular officials and managers, but recall that overt opposition was popularly regarded as out of the question. Those hostile to the management had to develop characteristics which may give the surface appearance of deference while concealing bitterness, equivalents of what Newby called 'the notorious ability of agricultural workers to touch their forelocks, while simultaneously raising two fingers behind their backs.'[75] It is not even clear that all miners did develop logical, coherent, and consistent views of the politics of the colliery villages and wider issues. Resigned acceptance is perhaps the most common response to powerlessness, and other topics exercise miners more than their 'working class self-image'. Just as in the case of workers in post-war Peterborough, the 'ideological dimension is of low salience in their images of society'.[76]

[73] Frank Parkin, *Class, Inequality and Political Order* (London, 1971), p. 84.

[74] Howard Newby, *The Deferential Worker* (London, 1977), p. 110.

[75] Newby, *The Deferential Worker*, p. 112.

[76] R. M. Blackburn and Michael Mann, 'Ideology in the non-skilled working class' in Bulmer (ed.), *Working Class Images*, p. 154.

We may therefore conclude by stressing once more that 'ideal type' sociological models of mining communities which would see left-wing political organizations as essential and automatic features of colliery-village society must be adjusted to take account of the circumstances and difficulties of the early stages of a coalfield community's development.

Table 27. *Southwell Rural District Councillors 1925-46*[a]

Ollerton		
M. F. M. Wright	1925-36	colliery manager
J. F. Jones	1925-37	landlord, the Hop Pole Hotel
W. S. Fletcher	1936-7	colliery manager
J. T. P. Foster	1937-46	landlord, White Hart Hotel
H. B. Watson	1937-46	colliery official
S. Kilner	1946-	Labour party, miner
G. L. Kirk	1946-	Labour party, miner
Edwinstowe		
P. W. Morley	1925-35	garage proprietor
G. E. Greaves	1925-39	Headmaster, Edwinstowe C. of E.
W. Tagg	1935-8	colliery undermanager
F. G. Gozzard	1938-46	checkweighman
H. Willett	1939-46	checkweighman
Mrs M. Beardsley	1946-	Labour party
F. C. H. Marsland	1946-	Labour party
Bilsthorpe		
L. T. Linley	1925-37	colliery manager
H. Barton	1937-46	colliery manager
C. Payton	1946-	Labour party, miner
Rufford		
F. S. Rolling	1925-40	farmer
W. V. Sheppard	1940-6	colliery manager
A. Francis	1946-	Labour party
Clipstone		
F. Armstrong	1925-45	solicitor
W. H. Mein	1925-8	colliery manager
G. I. Adkins	1928-34	colliery manager
W. Goodman	1934-5	colliery manager
W. A. Sansom	1936-42	colliery manager
S. Garner	1942-5	
T. E. B. Davis	1946-	Labour party
F. Clibbing	1946-	Labour party
Boughton		
Revd E. K. Hyslop	1925-37	Vicar of Boughton
Aaron Jenkins	1937-46	Spencer Union, checkweighman
F. Appleby	1946-	Labour party

[a] Sources: *Municipal Yearbook* 1925-46; local newspapers; Notts. CRO DC/SW 1/3/3; Southwell RDC Councillors signing-in book.

Table 28. *Nottinghamshire County Council elections 1925-46*[a]

Ollerton			
1925	Lord Savile	C.	unopposed
1928	Lord Savile	C.	unopposed
1931	Lord Savile	C.	unopposed
1934	M. F. M. Wright	Ind.	unopposed
1937	M. F. M. Wright	Ind.	unopposed
1938	F. E. Jones	Ind.	unopposed
1946	S. Kilner	Lab.	1,483
	J. T. P. Foster	Ind.	601
Bilsthorpe			
1937	Lady Savile	Ind.	763
	T. Todhunter	Lab.	389
1946	J. P. Todhunter	Lab.	1,018
	Earl Manvers	C.	496
Edwinstowe[b]			
1925	T. Warner Turner	Ind.	unopposed
1928	D. Warner Turner	Ind.	2,233
	O. Ford	Lab.	1,792
1931	D. Warner Turner	Ind.	2,055
	E. Poynter	Lab.	1,556
	J. Bennet	Ind.	236
	J. Ryan	Comm.	218
1934	E. Poynter	Lab.	2,850
	D. Warner Turner	Ind.	2,308
1937	W. M. E. Denison	Ind.	768
	G. G. Goodband	Lab.	267
1946	J. Reid	Lab.	1,175
	Lt.-Col. W. Denison	C.	601
Harworth			
1937	D. Buckley	Lab.	1,064
	W. Wright	Ind.	1,039
1946	J. Ainley	Lab.	unopposed
Clipstone			
1937	Mrs C. A. Taylor	Lab.	602
	C. Wilson	Ind.	428
1946	S. Clay	Lab.	unopposed

[a] Sources: local newspapers; B. F. L. Long, 'A study of the membership of selected local authorities', Nottingham University MA, 1964.

[b] In 1937, the ward boundaries were redrawn. Previously Edwinstowe had included part of the older coalfield near Worksop, which explains its election of a Labour councillor in 1934. In 1937 when the ward contained only the new village built for Thoresby Colliery, Labour did not win the seat.

Table 29. *Newark constituency,*[a] *parliamentary election results 1922–55*[b]

			%	electorate	turnout
1922					
Marquess of Titchfield	C	15,423	64.8	29,777	79.9
H. Nixon	Lab.	8,378	35.2		
C. majority		7,045	29.6		
1923					
*Marquess of Titchfield	C.	12,357	55.9	30,529	72.4
L. Priestley	L.	9,741	44.8		
C. majority		2,616	11.1		
1924					
*Marquess of Titchfield	C.	14,129	60.5	31,458	74.2
H. Varley	Lab.	5,076	21.8		
J. Haslam	L.	4,124	17.7		
C. majority		9,053	38.7		
1929					
*Marquess of Titchfield	U.	15,707	45.4	44,826	77.0
J. Haslam	L.	10,768	31.2		
W. R. G. Haywood	Lab.	8,060	23.3		
U. majority		4,939	14.3		
1931					
*Marquess of Titchfield	U.	25,445	70.1	47,788	75.9
Prof. J. R. Bellerby	Lab.	10,840	29.9		
U. majority		14,605	40.2		
1935					
*Marquess of Titchfield	U.	21,763	62.4	49,945	69.9
A. W. Sharman	Lab.	13,127	37.6		
U. majority		8,666	24.8		
By-election 8 June 1943 (elevation of Marquess of Titchfield as Duke of Portland)					
S. Shephard	C.	10,120	44.2	51,785	44.2
A. Daurant[c]	Ind. Prog.	7,110	31.1		
E. W. Moeran	CW	3,189	13.9		
J. T. Pepper	Ind. L.	2,473	10.8		
C. majority		3,010	13.1		
1945					
*S. Shephard	C.	18,580	45.1	56,447	73.0
H. V. Champion de Crespigny	Lab.	17,448	42.3		
H. F. Calladine	L.	5,175	12.6		
C. majority		1,132	2.8		

Table 29 (*cont.*)

			%	electorate	turnout
1950					
G. Deer	Lab.	28,959	54.2	60,660	88.2
*S. Shephard	C.	21,552	40.3		
E. H. Pickering	L.	2,950	5.5		
Lab. majority		7,407	13.9		
1951					
*G. Deer	Lab.	30,476	57.2	62,353	85.5
R. H. Watson	C.	22,817	42.8		
Lab. majority		7,659	14.4		
1955					
*G. Deer	Lab.	23,057	52.4	52,655	83.5
R. H. Watson	C.	20,916	47.6		
Lab. majority		2,141	4.8		

[a] Boundaries of Newark constituency: *1918–49*—Newark MB; Southwell, Newark, Bingham RDs; *1950–5*—Newark MB; Mansfield Woodhouse UD; Newark, Southwell RDs; *1955–70*— Newark MB; Newark, Southwell RDs.

[b] Sources: F. W. S. Craig, *British Parliamentary Election Results 1918–49* (1969), p. 446; *idem, British Parliamentary Election Results 1950–1970* (1971), p. 472; *idem, Boundaries of Parliamentary Constituencies 1885-1972* (1972), pp. 26, 83.

[c] In the 1943 by-election A. Daurant was supported by W. D. Kendall MP, his brother-in-law, and W. J. Brown MP. Kendall was elected Independent member for Grantham on 25 Mar. 1942, and claimed to be a member of the Labour party. Brown was elected Independent MP for Rugby on 29 Apr. 1942, receiving considerable support from local Labour activists.

* Incumbent.

Table 30. *Southwell RDC election results, 1946*[a]

Boughton			
F. Appleby	Lab.	301	elected
Revd E. K. Hyslop	Ind.	122	
*Aaron Jenkins	Ind	85	
(Labour gain)			
Clipstone			
T. E. B. Davis	Lab.	575	elected
F. Clibbery	Lab.	536	elected
Revd R. P. Wickens	Ind.	200	
H. W. Shaw	Ind.	180	
(two Labour gains)			

Table 30 (*cont.*)

Edwinstowe			
C. E. H. Marsland	Lab.	728	elected
Miriam Beardsley	Lab.	657	elected
F. Felstead	Ind.	264	
J. H. P. Morley	Ind.	222	
(two Labour gains)			
Rufford			
Annie Francis	Lab.	134	elected
*W. V. Sheppard	Ind.	121	
(Labour gain)			
Ollerton			
S. Kilner	Lab.	elected unopposed	
G. L. Kirk	Lab.	elected unopposed	
(two Labour gains)			
Bilsthorpe			
C. W. Payton	Lab.		elected
R. Brown	Ind.		
(Labour gain)			
Blidworth			
J. T. Brooks	Lab.	1,309	elected
W. Crewe	Lab.	1,162	elected
D. Darricott	Lab.	1,161	elected
J. Taylor	Ind.	431	
O. Blatherwick	Ind.	417	
T. B. Dodson	Ind.	280	
(three Labour gains)			
Worksop Rural District Council election results, 1946			
Harworth			
*A. Thompson	Ind.	1,205	elected
J. Smith	Lab.	1,104	elected
A. C. Slater	Lab.	1,100	elected
*Mrs M. K. Thomas	Ind.	1,075	elected
J. W. H. Brown	Ind.	872	
*H. Simons	Ind.	767	
(two Labour gains)			

[a] Sources: *Newark Advertiser*, 3 Apr. 1946; *Worksop Guardian*, 5 Apr. 1946.
* Incumbent.

Table 31. *Parish Council election results, 1946*[a]

Ollerton	
S. Kilner	656
G. L. Kirk	643
W. Swan	607
W. Carline	603
W. Charlesworth	567
F. Clarke	548
D. Ryan	539
A. Stockham	529
W. Dury	524
A. Knowles	496
J. Theobald	413
*F. Walster	276
*J. T. P. Foster	268
*W. Germany	263
*S. Thorneycroft	239
*B. Hunley	217
*W. H. Stockham	215
*F. Hutchinson	212
S. Brown	197

11 Labour elected
11 Labour gains

Bilsthorpe	
F. Pemberton	437
J. Metcalf	426
J. Litchfield	417
W. Hogg	416
H. Tuck	406
Mrs A. Luke	375
Mrs N. Topley	348
R. Stevens	328
*E. W. Lane	321
*G. Tordoff	239
*W. Hutchinson	222
*F. Wheatcroft	221
*A. Holmes	215
*R. W. Saresby	192

8 Labour elected
8 Labour gains

Edwinstowe	
D. C. Baines	807
*J. Reid	780
Mrs M. Beardsley	623
Mrs F. Reid	610
C. E. Marsland	572
T. Brocklebank	552
J. Barber	531
H. Marshall	511
N. Taylor	489
J. Bisam	486
W. Gent	445
N. J. Scothern	382
L. Davidson	363
*H. Jones	319
*J. H. P. Morley	313
*F. Felstead	309
*W. Trinder	285
*W. H. Russon	271
H. Gozzard	262
H. Jarvis	197

11 Labour elected
10 Labour gains

Boughton	
G. Ilett	313
F. Appleby	308
M. De Lacy	276
W. Foreman	261
J. Ryan	257
J. T. Young	255
G. Smith	246
A. Knowles	237
E. Stanley	194
*F. L. Beevers	187
*R. T. Thomas	184
*Revd E. K. Hyslop	160
*W. Wilson	151
*Aaron Jenkins	146
*T. Spray	137
*G. Hemingray	121
*S. Keeling	106
*G. Allcock	102

9 Labour elected
9 Labour gains

[a] Source: *Ollerton, Edwinstowe and Bilsthorpe Echo*, 5 Apr. 1946.
* Incumbent.

Table 32. *Electorates of Dukeries Mining Villages*[a]

Newark constituency:

	Ollerton	Boughton	Edwinstowe	Bilsthorpe	Clipstone
1920	334	133	441	64	171
1921	335	136	441	69	249
1922	343	171	470	67	304
1923	354	179	461	73	335
1924	502	186	488	78	401
1925	650	197	552	87	697
1926	805	278	707	146	1,086
1927	1,512	313	727	230	1,294
1928	1,689	375	769	388	1,562
1929	2,117	435	840	714	1,875
1930	2,152	578	1,459	999	1,884
1931	2,303	721	1,642	1,033	1,863
1932	2,292	721	1,876	1,053	1,901
1933	2,264	736	1,891	1,035	1,873
1934	2,338	749	1,878	1,033	1,869
1937	2,182	615	1,883	999	1,803

Mansfield constituency:

	Blidworth
1921	560
1922	605
1923	629
1924	678
1925	809
1926	1,138
1927	1,681
1928	1,789
1929	1,919
1930	1,873
1931	2,086
1932	2,226
1933	2,344
1934	2,614
1935	2,858
1936	2,838
1937	2,781

[a] Newark and Mansfield Electoral Registers, 1920-37.

7. Education

'This school has been subject to almost constant difficulties'[1]

When the exploitation of the concealed coalfield east of Mansfield in the 1920s led to the foundation and rapid expansion of colliery villages to serve the new pits, one of the services which had to be provided for the new communities was education. In order to cope with the influx of immigrant miners, many of them with children, new schools had to be built. The existing villages' church, voluntary, and council schools were quite unfitted for the task of educating many hundreds of newcomers in each community. The history of the provision of schooling in the new villages demonstrates once more how the impact of mining in a previously backward agricultural district caused dislocation and dissatisfaction, and how a conflicting tangle of authorities led to a serious gap between the arrival of the immigrants and the provision of essential services. Education fits into the pattern of the unhappy early history of the new coalfield communities.

First, let us consider the steps taken by the Nottinghamshire County Council, which was responsible for building the new schools made necessary by the growth of population. In April 1924 the Bolsover Company put forward plans for the erection of 956 colliery houses in Edwinstowe parish.[2] In the same month, B. W. L. Bulkeley, the Secretary to Nottinghamshire County Council's Education Committee, submitted a scheme for the building of a new public elementary school for 1,000 children in Edwinstowe.[3] A site of 5.389 acres was purchased from the Bolsover Company at a cost of £105 per acre.[4] When central government suggested that it was too

[1] PRO ED 21/37593, School Inspector's Report, Harworth Bircotes Council School, 14 May 1929.

[2] *Mansfield Reporter*, *Sutton Times*, 18 Apr. 1924.

[3] *Mansfield and North Notts Advertiser*, 18 Apr. 1924.

[4] PRO ED 21/37571, Edwinstowe Council School.

large, Bulkeley defended the size of this site: 'As the Board are aware, a large industrial population will have to be accommodated in this neighbourhood . . . and it is difficult to say what the ultimate sizes of the mining villages will be.'[5] In June 1924 the cost of the new school was estimated at £21,650. Water supply and sewerage systems were to be provided by the coal company. However, the uncertainty current during the early stages of the development of the new coalfield then took a hand. In 1925 Bolsover decided to erect only 100 houses immediately since coal had not yet been reached, and in November of that year the Education Committee decided to proceed with only one wing of the new school, to accommodate 352 children. In the meantime an army hut was to be imported from Clipstone Camp and outside toilets were to be brought from Rainworth to form a temporary school, which was opened on 12 April 1926.[6] The Head Teacher was John Henry Land, aged only 22. A tender of £11,985 was accepted for the building of the first stage of the permanent school, which was opened on 31 August 1927. The temporary hut was then no longer in use and an older Headmaster was appointed. As numbers in the village continued to grow, it again proved necessary to call the army hut into use in September 1930, and the enlargement of the school at a cost of £7,995 was now proposed. The new wing was completed in December 1931.

At Bilsthorpe too, the development by the Stanton Ironworks Company had persuaded the Education Committee to announce the construction of a new council school in September 1924,[7] but it proved necessary to press into use one of the ubiquitous ninety-six-seater huts in June 1925, and the school expanded into the village's Mission Hall in July 1927. The permanent school was not opened until 4 June 1928, at a cost of £22,000; by this time 255 pupils were on the roll.[8] This increased to 411 by October 1929, although as late as 1932 the school was still registering eighty-eight new admissions and ninety-two leavers *during* the school year, a testimony to the fact that the population of Bilsthorpe still

[5] Ibid. [6] Ibid.
[7] *Nottingham Journal, Nottingham Guardian*, 6 Sept. 1924.
[8] PRO ED 21/37496, Bilsthorpe Council School.

had not stabilized. At Harworth, the largest of the new mining villages, it was found necessary to open three new schools on the same site, in 1923,[9] 1925, and 1929.

At Langold, a variety of temporary accommodation had to be provided before the permanent school was ready for occupation. In 1923 one of the itinerant huts which proved so useful to the Nottinghamshire Education Committee in these years of colliery development was pressed into service. As Bulkeley informed the Secretary of the Board of Education: 'On this site my Committee propose to erect an iron building accommodating 120 children. This iron building was originally erected in 1906 on the Forest Town (previously known as New Sherwood) site, Mansfield Woodhouse, and is now disused.'[10] It was of course necessary to put up a permanent building, plans for which were approved on 1 September 1924; the cost was to be £20,669. 13*s.* 1*d.* Nevertheless, before this school could be occupied, further emergency measures had to be taken by the education authorities. In 1925, a second hut was brought to Langold, and a more unusual expedient was adopted in 1926. Bulkeley explained the problem to the Secretary of the Board as follows:

> The two temporary buildings, accommodating 220 scholars, are already overcrowded, and in addition 70 children are being conveyed daily by bus to the Voluntary School in the adjoining parish of Carlton-in-Lindrick. There are a few children resident in Langold who do not attend either School, but there are a further 100 colliery houses ready for occupation and these will be let within a few weeks' time. Thus there will be a considerable child population unprovided for. My committee propose, subject to the approval of the Board, to rent the first floor room in the new premises recently erected by the Worksop Co-Operative Society at Langold, near to the present school buildings.[11]

After the room above the Co-operative store was first used on Monday 12 April 1926, the scattered Langold council school had a capacity of 310 pupils. When five classrooms of the permanent school were opened on Wednesday 1 September 1926, the Co-op hall could be vacated, but the huts could not

[9] PRO ED 21/37593, Harworth Bircotes Council School.

[10] PRO ED 21/37062, Hodsock Langold Council School, Bulkeley to the Secretary of the Board of Education, 30 Oct. 1923.

[11] Bulkeley to the Secretary of the Board of Education, 10 March 1926. PRO ED 21/3/37602.

be dispensed with until 10 January 1927, when the rest of the new school was completed. However, by this time events had already overtaken the harassed Education Committee. Under the headline 'Mushroom Villages–Second Notts School Needed Ere First Completed', the *Nottingham Evening Post* of 2 November 1926 had reported that

> The amazing manner in which new colliery villages spring up in North Notts was illustrated at today's meeting of the Notts County Council.
>
> Colonel H. Mellish mentioned that part of the new school had been completed and occupied at Langold, and the remainder would be ready after Christmas. By Christmas, however, there would be 800 houses in that village, and more than 1,000 children, so that the 700 odd places in the school would be more than occupied. The Education Committee had therefore to face at once the problem of building a second school on that site.

The County Architect was instructed to prepare plans for an infant school to be situated on the same site as the new council school. This opened on Monday 3 September 1928 at a cost of £10,030.

Finally, at Ollerton the Whinney Lane council school, notified in September 1924,[12] was not opened until September 1927. The final part of the delay was due to the failure of the Southwell Rural District Council to complete the connections of the water and sewerage systems to the school, or to metal Whinney Lane itself. This led in the summer of 1927 to a row in the columns of the local press between the Chairman of the County Education Committee, Alderman Edge, who claimed that the Southwell RDC was guilty of negligence, and the Chairman of the Rural District Council. As will be seen below,[13] the Butterley Coal Company also voiced its displeasure at the inefficiency of the rural district council. As usual, temporary accommodation had to be provided. It took the form of Wellow Church Hall and the New Ollerton Mission Room, but with the opening of the permanent school in 1927 a capacity of 728 was attained, which had been increased to 1,100 by July 1929 with the addition of a separate infants' department.[14] A second new council school in Ollerton, at

[12] *Mansfield Reporter, Mansfield and North Notts Advertiser,* 29 Sept. 1924.
[13] See below, pp. 172–3.
[14] PRO ED 21/37659, Ollerton Whinney Lane Council School.

Walesby Lane, was notified on 28 May 1931 and recognized as efficient in August 1934; 274 children could be accommodated there.[15]

How good an education did these new schools provide? Here the best evidence is provided by central government archives, the Schools Inspectors Reports in the Public Record Office.[16] In general, these reports reveal that serious problems were posed by the development of the new coalfield in the Dukeries, particularly as a result of the high level of immigration and labour circulation of miners that this entailed. The inspectors visited the Edwinstowe Council school on 25 and 31 May 1927:

> This school had been in existence for just over a year and is supplying the needs of a growing part of the parish, consequent upon the development of the local coal field. The children are mainly migrants from other parts of the country and from neighbouring parts of Derbyshire. The teaching and work of the Senior Department are still in a state of flux and the children are not making much headway. The Head Master is young and very inexperienced and though willing and industrious is finding his duties too difficult for his present powers.[17]

Five years later, with a new and older Headmaster, the problems were still there:

> Built to serve a new colliery area, the School has in the last year or two grown very rapidly. Its admissions registers, however, like those of other Schools of this kind in the County, speak of the vicissitudes and uncertainty of existing industrial conditions: since April 1st last, for example, 163 children were withdrawn from the school, their parents leaving the district; in the same period there were 154 new admissions.[18]

Similarly unsettled conditions were found at Blidworth. In September 1928, the Inspectors reported that

> the children have been drawn from as many as eight different counties; some have suffered from the frequent migration of their parents, a certain number are of poor physical type and a few show mental defects . . . the difficulties of the school have been formidable and it must be some time before it can be judged from the point of view of schools working under happier and more settled conditions.[19]

[15] PRO ED 21/37657, Ollerton Walesby Lane School.
[16] PRO ED 21.
[17] PRO ED 21/37571.
[18] PRO ED 21/37571, 4–5 Feb. 1931.
[19] PRO ED 21/37501, 6–7 Sept. 1928.

Yet in the report of an inspection made on 30 September 1931, we find that

the difficult circumstances of the school, noted in the report of September 1928, still persist. Children come and go as their fathers find or fall out of employment in the coal mining area the school serves. The names of many pupils appear, re-appear and appear again in the admissions register. Of 136 new admissions since April 1st last only 53 entered upon reaching school age; the other 83, older children from widely separated districts and schools, were distributed according to age through the various classes. Standard I, at the time of the inspection, had 54 children on its register . . . only 18 of them had been in continuous attendance since the beginning of their school lives. Pupils subject to such vicissitudes can scarcely be expected, whatever their nature [*sic*] ability, to reach normal standards of attainment.

Between 1927 and 1932 there were 1,100 admissions to the school, and between August 1932 and March 1934 still a further 120 were admitted, seventy-seven being migrants, while during this period seventy children left owing to their parents' removal.[20]

At Langold in 1932 the Inspectors picked out the difficulties of children migrating from other areas as one of the serious problems affecting pupils in the school's 'C' stream:

The slow-movers in the 'C' sections are nearly all the victims of fortuitous circumstances, having come from schools in other districts, or at a later age because of ill health. Some of them will make good before they reach the Senior School, but of others it is doubtful if they will ever attain a very sound academic standing.[21]

A study of the admissions register of Langold's Temporary School indicates some of the disruptive circumstances the teachers had to contend with. The initial roll of the school in 1924 numbered eighty-seven children, but 148 entered in 1925, 283 in 1926, and a peak of 314 in 1927. The figure settled down at between 157 and 179 in the period from 1928 to 1931. Children were coming and going throughout the year. Of the 1,891 pupils who attended the school between 1924 and 1932, 576 left the district before completing their education at Langold school. Some even returned to Langold for a second stay, 139 of the 1,891 admissions being readmissions.[22]

[20] PRO ED 21/37501, 15 Jan. 1932, 20 Mar. 1934.
[21] PRO ED 21/37602, Inspector's Report, 28 June 1932.
[22] Notts. CRO SA 105/1/1.

At Harworth Bircotes Council School, there was a 'thoughtful, vigorous and enthusiastic Head Master who is deeply interested, from his former experience, in the problems of a mining village.'[23] This was fortunate, because the problems were great. The Inspectors reported in May 1929 that

> This school has been subject to constant difficulties arising from the migration of the parents . . . during the period of twelve months immediately preceding the inspection, 370 children had left the school, the number at present on the books totalling 419. Such conditions give the Head Master and staff very little chance of stabilising their work.[24]

As late as December 1931, still

> hardly anything is consistent in this school; there is a large amount of migration each year and the number of those who go through the full infants course is small . . . generally under these conditions, it is not surprising to find that the standard of work is somewhat low.[25]

An interviewee who came to Ollerton from Warsop just after the pit had become productive found his education similarly disrupted:

RJW Now presumably you would have been of school age when you came to Ollerton?

CB Yes, I was seven.

RJW Did you have any trouble getting to join a school in Ollerton?

CB Yes, indeed. We arrived in November of '26 and first of all Mother took us—that's myself and my brother—down to Old Ollerton, to the National School at Wellow Road, and we couldn't get in there. In fact, Mr Greaves, the headmaster, told us that there was a waiting list of no end, and there was no hope of us getting in there in the foreseeable future. And then Mother took us to the Church School at Boughton. My brother got in because there happened to be some vacancies at his age—he was three years older than me—but I didn't, so I just didn't have any schooling until they opened the one at Whinney Lane. I don't know what precise date they opened that, but it was the following year, wasn't it?

RJW Yes. So you had a break of a few months before you could go to school again—you had been to school in Warsop?

CB Yes, yes, and actually my mother asked the teacher at Warsop if we could have a few exercise books, because she knew what the situation was at Ollerton, and they were just incredulous and they

[23] PRO ED 21/37593, 14 May 1929.
[24] Ibid.
[25] PRO ED 21/37593, 15 Dec. 1931.

said 'no, they're bound to have schooling there', you know, you couldn't escape schooling, but we were just left and ignored, nobody bothered. I suppose these days they'd probably have bussed us somewhere. But nobody bothered, we just stayed on holiday.

The problems were not over once one had registered at a school:

RJW How good was the schooling at Whinney Lane?
CB Well, I think the buildings were finished, but I do remember the first day or two we actually sat on the floor because they didn't have any desks.[26]

Nor were the older schools in the villages able to offer a better service in the face of the influx of miners' children. In 1925 the Nottinghamshire County Council's blacklist of schools with inadequate accommodation included Blidworth C. of E., Blidworth Wesleyan, Boughton C. of E., Edwinstowe C. of E., Harworth C. of E., and Ollerton C. of E.[27] Faults reported included overcrowding, and inadequate lighting, heating, ventilation, toilets, claokrooms, and playgrounds. The Inspectors wrote in 1925 of Edwinstowe Voluntary School, which was not even on the blacklist, that

There is much passage room difficulty. Ventilation is very bad indeed, and the heating is insufficient. Offices are too close to the school. The cloakroom accommodation is insufficient. Playground is not suitably paved. Structurally this school is very difficult to remedy.[28]

The old Bilsthorpe council school was seriously overcrowded as a result of new entrants, who came from a number of counties, showed 'mixed attainments', and often had 'broken school careers'.[29] This school was closed in 1928 on the opening of the new council school. At Blidworth, the old Church of England school was severely criticized for the standard of both its teaching and its premises in the Inspector's Report of 27 August 1926.[30]

[26] Interview, Cyril Buxton, Forest Town (Ollerton).
[27] PRO ED 21/99 Nottinghamshire Premises Survey, 1925.
[28] PRO ED 21/37570 Edwinstowe Voluntary School. This school was privately owned and part of the Manvers Estate.
[29] PRO ED 21/37495, Bilsthorpe Council School.
[30] PRO ED 21/37500, Blidworth C. of E. School.

The new schools built in the 1920s at least avoided most of these material shortcomings once they *were* opened, but apart from resembling building sites during the lengthy period before they were completed, they still suffered thereafter from poor resources and frequent staff changes. At Bilsthorpe Council School there was an almost complete change of staff at each two-yearly visit of the County Education Committee: the turnover was so great that new paper had to be pasted over the 'Staff' column in their report ledgers.[31]

It is scarcely surprising that the coal companies were concerned lest the inadequate schooling might deter men with families from coming to work at the new pits. In the spring of 1927, for example, the Butterley Company's mining agent Eustace Mitton complained about the non-opening of Ollerton Whinney Lane Council School: 'The delay of the Southwell RDC in all their matters affecting roads, sewers, water supply etc. is beyond comprehension.'[32] However the blame for the delay should not be laid solely at the door of the Rural District Council. When on 18 December 1925 Ollerton Colliery Manager Montagu Wright wrote to the County Education Committee asking about educational facilities, since the miners coming to the village would be worried as all the Ollerton schools were now full, the Committee's Secretary, B. W. L. Bulkeley, replied: 'I am afraid we shall be very much behindhand with the school. I had no idea that you are going to proceed at such a pace. I wish you could have let us know a little sooner . . . in fact the delay is with the Board of Education. The plans have been there now for nearly two months.'[33] On 2 March 1926, Mitton wrote to Bulkeley:

I look with anxious eyes every time I go to Ollerton to see the new school rising on the Site which was arranged between the Butterley Co., Lord Savile and yourselves, and I am somewhat surprised and perplexed by non-progress. You are aware that the population is rapidly increasing in this District, and I understand that children are getting very crowded in the Schools, and what is puzzling me very much is that

[31] Notts. CRO ED/CC/8/10/31.
[32] Butterley H3 222, 10 Mar. 1927.
[33] Butterley H3 222, 23 Dec. 1925.

the manager, Mr. Montagu Wright, tells me that you have built a school at Edwinstowe[34] before the Bolsover Co. have even started to sink, whereas at Ollerton the seam has been proved, but you have taken no further steps in the matter.[35]

Lack of communication and co-operation between the coal companies and the authorities responsible for facing the problems consequent upon the development of the coalfield led to an inadequate education service being provided in the early years of the new villages, just as roads, water supply, and sewage systems had lagged behind the arrival of the mining population.

In many cases the staffs of both old and new schools in the Dukeries mining villages struggled manfully to provide a reasonable elementary education for the children who passed through their hands. But compared to the schools in older mining areas such as those around Hucknall and Eastwood in west Nottinghamshire,[36] the schools in the new villages faced very different problems: overcrowding and turnover rather than the difficulties of teaching in outmoded Victorian buildings. Virtually none of the children living in the new coalfield received more than a minimum elementary education or passed on to the nearest grammar schools, in Mansfield.

The impression of segregation between the children of the old village families and the miners' children was sometimes reinforced by the long-standing connections of the former with the old church schools in the villages. Nor did many of the Dukeries colliery village children benefit from long-established scholarship schemes of the miners' union. It is clear that those few scholarships which were available did not cover the whole cost of further education, and it was difficult for most miners to make up the sum required.

There was scarcely any attempt at adult education in the new colliery villages, like that provided by the Workers' Educational Association in established coalfields such as that of Derbyshire.[37] The only form of education provided apart

[34] Actually only a temporary wooden hut.

[35] Butterley H3 222.

[36] For example PRO ED 21/37605, Hucknall Beardall Street Council School; PRO ED 21/37563, Eastwood Council School; PRO ED 21/25805, Heanor Loscoe Rd. Council School (Derbyshire).

[37] J. E. Williams, *The Derbyshire Miners* (London, 1962), pp. 471–2, 792–3.

from the elementary schools consisted of evening classes in mining and first aid. At Harworth the school premises were used in the evening for compulsory safety classes and mining science, taught by the undermanager of the colliery.[38]

It is fair to conclude on this practical note that education provides another example of the serious dislocation characteristic of the early years of the new Nottinghamshire coal-mining villages, and of their social disadvantages, which exceeded even those to be found in Britain's more established coalfields.

Table 33. *New schools in the Dukeries mining villages*[a]

	notified	opened	cost (£)	capacity
Edwinstowe Council	10 Apr. 1924	Aug. 1927	12,264	352
		Dec. 1931	7,995	
Bilsthorpe Council	30 Aug. 1924	Jan. 1928	21,566	792
Clipstone Council	29 June 1923	Aug. 1926	15,550	728
Blidworth Council	10 Apr. 1924	Sept. 1927	21,517	744
Harworth Bircotes Council	25 Jan. 1921	Sept. 1927	5,912	288
	23 Nov. 1923	Sept. 1925	10,900	728
	July 1927	Aug. 1929	8,981	1,016
Ollerton Whinney Lane	29 Sept. 1924	Sept. 1927	21,521	712
	Sept. 1928	July 1929	8,823	1,100
Ollerton Walesby Lane	28 May 1931	Aug. 1934	6,090	274
Hodsock Langold Council	31 July 1923	Jan. 1927	20,670	744
	12 Nov. 1926	Sept. 1928	10,030	388

[a] Source: PRO, Schools Inspector's Reports.

[38] Interview, Stan Morris, Harworth.

8. Religion

'They owned you body and soul as well . . .'[1]

The private coal companies which planned and built the new villages of the Dukeries field were concerned with many aspects of the lives of the inhabitants. Not only did they regulate matters directly within their jurisdiction, such as conduct on company property in the pits and villages, but they were well aware that in order to protect their interests they would have to monitor affairs technically in the province of some other authority. For example, they were concerned when the County Council proved dilatory in providing for the schooling of the children of the miners who came to work in the new pits.[2] Religious activity in the colliery villages was no exception as far as the companies' policy of intervention was concerned. Indeed, in many of the communities the coal company actually took at least partial responsibility for the building of churches and chapels for the miners.

This is not to say, of course, that the coal companies alone were interested in 'bringing religion' to the Dukeries mining villages. As will be seen below, it is clear that both within and outside the new communities active members of a number of denominations recognized that the development of mining in the area constituted a challenging change of circumstances. As far back as 1910 the problems of religious institutions at a time of social, demographic, and economic transformation were pointed out by the Bishop of Southwell in a pastoral letter:

Our task in this diocese is great. The two counties of Derbyshire and Nottinghamshire contain the vast Midland coalfields. Year by year there is new development. Country villages change suddenly into colliery villages, where churches, or mission rooms are needed. In such parishes

[1] Interview, J. H. Spencer, Ollerton.

[2] See above, pp. 172–3.

often times the staff of the clergy is wholly inadequate, and the mass of the people, ignorant and unshepherded, are tempted to drift into the ranks of the indifferent or hostile, save where other religious bodies assert their influence.[3]

The Wesleyan Methodists of Blidworth also identified the new challenges. On the opening of a new chapel in the village in 1932, R. P. Blatherwick wrote:

It is hardly for me to write about the men and women, who today carry on the work, and are in a line of great succession. The House of God stands as a constant reminder of the unchanging amid the changing. Man's physical and material environment changes with the years, but his urgent needs are unalterable, and can only be met by Him in whose Name the Sanctuary is built.[4]

Nevertheless, the coal companies did play a leading role in finding solutions to the problems of founding churches in the model communities. As usual, the energetic coal-owners of the Dukeries did not do things by halves. The Butterley Company decided to build a church at the geographical centre of New Ollerton colliery village as 'a cathedral for the new coalfield'. If this was to be done, it was to be done properly. On 16 April 1926 mining agent Eustace Mitton wrote to Sir Giles Gilbert Scott, architect of Liverpool Cathedral, asking him to submit plans for a church and vicarage at Ollerton.[5] On 9 July, Sir Giles was brought by company car to survey the site, Church Circle, the focal point and centre of New Ollerton colliery village. Scott agreed to submit plans, but the company compounded their boldness in approaching him by rejecting his designs and dismissing him as architect. In the end the church was designed by Messrs. Naylor, Sale, and Woore of Derby and built by Messrs. Greenwood of Mansfield, at a total cost of £8,000, to which the Butterley Company contributed £5,000.[6] The C. of E. church of St. Paulinus was consecrated on 1 October 1932. As the *Southwell Diocesan Magazine* put it: 'The whole

[3] *Derbyshire Times*, 30 Apr. 1910, reported in J. E. Williams, *The Derbyshire Miners* (London, 1962), p. 466.

[4] R. P. Blatherwick, Methodism in Blidworth (1932), unpublished, in Methodist Archives, John Rylands University Library of Manchester.

[5] Butterley L1/1 227.

[6] Butterley F1/4 202.

Diocese will rejoice that a Church has been built in their colliery district to meet the spiritual needs of the new population.'[7]

Spiritual needs perhaps; yet the curious fact about the company's generosity was that the branch of the Wright family which controlled Ollerton Colliery was staunchly Roman Catholic, and the agent, Mitton, had to be sent to the ceremony of consecration as the company's representative.[8] Perhaps the church was also seen as an investment which might reap rewards in the form of the social control as well as the spiritual guidance of the Ollerton miners. The company did at least regard themselves as benevolent paternalists, and they also aided financially the Baptist, Methodist, and, of course, Roman Catholic churches in Ollerton, along with the Salvation Army.[9]

Nevertheless, it was felt by some that some churches were more favoured than others. It was rumoured in Ollerton that men might do well to change their religion to suit the management:

CHG And then of course they built a Catholic church, because in the old days if you were a Catholic at this pit, you got on pretty well.
RJW Why was that?
CHG Because the Manager was Catholic. He was one of the Wrights of Ripley. I don't know whether you've heard of that family.
RJW Yes.
CHG Butterley Company. In fact, I do believe that all Catholics had a shilling a week stopped, for that church.
RJW Were there quite a lot of Catholics in Ollerton?
CHG Yes, but there were quite a few changed to be Catholics![10]

Another interviewee claimed personal experience of this phenomenon:

But then there was this element, which I think we talked about before,

[7] *Southwell Diocesan Magazine*, Sept. 1932, p. 149.

[8] Disraeli's 'Mr Trafford of Wodgate' was another Roman Catholic employer who endowed Anglican churches: 'in every street there was a well; behind the factory were the public baths; the schools were under the direction of the perpetual curate of the church, which Mr. Trafford, though a Roman Catholic, had raised and endowed.' (B. Disraeli, *Sybil*, 1845, Bradenham Edition, p. 212.)

[9] Butterley S6 311. Montagu Wright's mother, Mrs Wright of Staines, gave £1,200 towards the total cost of £3,000 needed to build Ollerton's Roman Catholic church. (*Newark Advertiser*, 12 Sept. 1928.)

[10] Interview, C. H. Green, Ollerton.

of Catholicism: all the bosses seemed to be Catholic. And I know of one case certainly where this person changed his religion in order to be well in so to speak, and he's somebody very close to me too! So I know it's true! I remember almost a family crisis at home when my brother was getting married, because he was marrying the daughter of the person who'd been converted to Roman Catholicism, you see, and therefore the wedding had to be in a Roman Catholic church, and my grandfather knew all the forebears of these people: 'They're not bloody Roman Catholics, what are they getting married there for!', you see, and this went on for weeks. 'He only bloody changed his religion!' Actually he had some dogs, and when Monty[11] used to go shooting he used to take his receivers and what not, he was a shot himself you see and 'he's only following the bloody full cart, that's why he's changing to Roman Catholic!' This sort of thing going on.[12]

Further evidence is to be found in the archives of the Butterley Company itself. On 11 July 1929, the Vicar of Ollerton, E. F. H. Dunnicliff, wrote to mining agent H. E. Mitton to complain that all the miners at Ollerton Colliery had had 1*s.* tickets for the sports day in aid of the Roman Catholic church attached to their lamps at the pit. If the tickets were not returned immediately the money would be deducted from their wages. Dunnicliff further complained that: 'Without being unduly swayed by prejudice, I am beginning to fear that to be a Churchman at Ollerton pit, is to be liable to dismissal upon slighter grounds than, shall we say, to be a Roman Catholic.'[13]

Coal companies built churches elsewhere in the new coalfield besides Ollerton, at Bilsthorpe, Harworth, Clipstone, and Langold, though not at Blidworth or Edwinstowe, where the old village churches were more easily accessible.[14] At Clipstone, the foundation stone for the Anglican All Saints Church was laid on 22 April 1928 by the Duke of Portland, who spoke of:

the remarkable developments taking place in the North Notts coalfield, and the astonishing increase of population brought about in consequence . . . His Grace recalled the time only a few years ago, when Clipstone was a wide, open moor, whereas today there is a

[11] Montagu Wright, Ollerton colliery manager.
[12] Interview, Cyril Buxton, Forest Town (Ollerton).
[13] Butterley H1 418.
[14] See Table 34.

resident population of 4,000, with every prospect of another 2,000 being added in the course of a short time.[15]

About £8,000 was needed for the new church, All Saints, which was designed to meet the needs of the new population. The Bolsover Company gave the site and £3,000, the Duke of Portland, the Marquess of Titchfield, and the Ecclesiastical Commissioners gave £1,000 each. A contract for a church without tower and spire costing £6,000 only[16] was let to Messrs. J. F. Booth and Son of Banbury, builders, and Louis Ambler, architect. Before 1928 church services were held in an old army hut, but the new church would seat 375. The Revd Day Lewis, Vicar of Edwinstowe, in whose parish New Clipstone was then sited, commented: 'They might build up the Empire by buying British goods, but they could only build up the people in the faith of Jesus Christ by having a church in their very midst.'[17] In 1930 a new and separate ecclesiastical parish was created for Clipstone.[18]

At Bilsthorpe, there was a rumour of dissent between the management and the church in the old village:

The manager at this colliery fell out with the vicar[19] and he decided that he was going to have his own church, and successfully did it. He got another clergyman, they had a little church built, where the main Welfare is now . . . any road eventually this church was built, and the manager, of course, he'd got a hell of a following. You know for you to get on in a pit in those days you'd got to be well in with t'gaffer, kind of thing, and he'd got a lot of people that, if he said bow they bowed and that, purposely for money, at the end of the road it were all money. And so therefore of course, all the churchgoers in those early days who went to St. Margaret's Church, followed the manager and they went to the new church.

However, the company was not quite so eager to build a Methodist chapel in Bilsthorpe:

RJW What about chapels, Methodist or anything like that? Were many miners Nonconformists, or Methodists?

HT Oh, a lot of Methodists. But it was many years before it was asserted, and I think this was brought about once more by the

[15] *Worksop Guardian*, 27 Apr. 1928.
[16] The church was never completed.
[17] *Worksop Guardian*, 27 Apr. 1928.
[18] *Newark Advertiser*, 27 Aug. 1930.
[19] For a possible reason for this quarrel, see below, p. 183

> gaffer of this pit being Church of England. You see, you had to be very careful what you set up in a colliery village, to oppose management, very careful indeed, because of your jobs. But eventually, a long, long while after, I'd say at least twenty years after the pit was sunk, they eventually got a Methodist church.[20]

Curiously, this reflects earlier difficulties that the Methodists had found with the local powers that be at Bilsthorpe: in 1893 Lord Savile had refused to sell land for a Wesleyan chapel at Bilsthorpe.[21]

The Church of England in Harworth was also encouraged by the management:

> Another thing about attending church in the old days, generally if the colliery manager was connected with the church, like in the old days in the wool mills, the mill owners in Yorkshire used to sort of build the chapels, and everybody had to attend. Well, it was much similar when we came to Harworth, the people attended because the colliery manager attended there, and I think it was mainly to gain favour, you had to keep well in with the manager, he attended there and all the colliery overmen and deputies used to attend there, so they had really a very good congregation . . . the Church of England had got their building on Whitehouse Road, which was provided wholly by the Barber-Walker Company, they had built them the place, they had built the vicarage, they got the land, they provided a certain amount of money for the stipend of the vicar, they provided him with coal and basically ran the church for him.[22]

Apart from the building of new churches, the coal companies showed in other ways that they were interested in the 'spiritual well-being' of their miners. Butterley's mining agent Eustace Mitton and Lord Savile's agent John Baker had combined to finance the work of a lay reader and Church Army Missioner for the sinkers at Ollerton. These clerics had worked from the cemetery chapel of Old Ollerton Church from as early as 1923.[23] A temporary Mission Room was then provided by Butterley for church services between 1926 and 1932 when St. Paulinus was opened. Butterley paid £80 towards the New Ollerton curate's stipend of £350 per annum, and their contribution was matched by Lord Savile. The rest of the money was found by the Ecclesiastical

[20] Interview, H. Tuck, Bilsthorpe.
[21] Notts. CRO MR 5/212, J. A. Bell to the Revd R. Nicholson, 15 Mar. 1893.
[22] Interview, Stan Morris, Harworth.
[23] Butterley E3/4 145.

Commissioners (£60), the Church Extension Society (£30), and the Additional Curates Society (£100).

Butterley felt that this financial interest gave them a role in deciding who the prospective curate for the colliery village should be, and several candidates were interviewed by the Company Welfare Director, Colonel Banks. The first candidate seemed most suitable: 'A great Sportsman, a great Football man, and has a very good recommend',[24] wrote mining agent Mitton. Unfortunately this clergyman required too much money to come to Ollerton. The next candidate, a Revd Boykett, was rejected because of his lack of interest in the Boys Brigade. Butterley and Savile then approached H. W. Good of Beeston, who to their surprise preferred to become Bishop of Madagascar rather than come to Ollerton. After another failure, the Revd FitzHugh (described by Colonel Banks as 'a bit of a thruster'), in 1926 the ideal man was found: the Revd Ross, a former captain of soccer at Rugby School who at interview was found to wear a moustache (which meant that he was not a High Churchman).[25]

When reading the Butterley Company's file on the appointment of a curate for New Ollerton, it is difficult to believe that in effect the choice did not lie entirely in their hands. The investigations and interviews that they conducted could have been concerned with the appointment of any middle-ranking company official. Mitton certainly felt a degree of responsibility for the provision of religion for what he called 'the District that my company has brought into being'. He wrote to the newly appointed Ross on 10 November 1926 that the Church must be 'a Leader for the great coalfield now opening up in Nottinghamshire'. Butterley maintained an interest in the personnel involved in the leadership of the Church of England in Ollerton throughout the inter-war period. In 1932 there was a vacancy in the New Ollerton curacy. The Vicar of Ollerton, E. F. H. Dunnicliff, consulted closely with the Butterley Company management at the colliery. He reported, 'I have prospects of getting a good man, as the two at present in view are Varsity men who are good at games, and interested in working people.'[26]

[24] Ibid., U1 419.
[25] Ibid., Banks to Mitton, 11 June 1926.
[26] Ibid., H3 222.

The interest in the character of the local clergyman extended to the vicar of the Parish of Ollerton himself. When the Revd S. J. Galloway[27] succeeded the Revd Dunnicliff[28] in April 1934, John Baker was reported as saying that 'Ollerton was very fortunate in having such a splendid Sportsman as the Rev. Galloway undoubtedly was in his opinion a great asset, as he considered that for a clergyman to be a good Sportsman, meant that he must also be a good Pastor for his Flock.'[29] This is one more example of how the dominant authorities in the Dukeries were aware of the importance of pastoral guidance in religious and sporting matters alike.[30] This version of the idea of 'muscular' Christianity was also to be found at Harworth: 'He was a good curate, he who came here,[31] there can be no doubt. He could swim, box, play rugby, football, cricket—I've never seen a chap who could knock fifty as quick as he could.'[32]

The relationship between the coal company and the local clergyman was not always harmonious. In 1932 an independent Anglican clergyman was appointed to work among the miners in Bilsthorpe colliery village. The Stanton Company Secretary, A. E. G. Harmon, later wrote:

> The development of Bilsthorpe Colliery and attendant amenities for employees gave rise to consideration of the appointment of an anglican clergyman to work independently among the villagers—a proposal with which the bishop of Southwell expressed sympathy. In 1929 the directors offered to contribute £240 annually towards the stipend of a cleric, providing the Ecclesiastical Commissioners would add £60, making £300 in all, but the subject appears to have fallen into abeyance.

[27] Revd Sydney John Galloway, MA. Born Papworth Everard, Cambridgeshire, 1886. Ed. Haileybury and Jesus College, Cambridge. M. 1934 Joan, daughter of Major T. P. Barber, DSO. Vicar of Greasley, Notts., 1920-33. Vicar of Ollerton 1934. Army Chaplain 1915-20. President, Notts. County Hockey Association, 1933-4; played hockey for Notts. County. Recreation—games. (*Who's Who in Nottinghamshire*, Worcester 1935, p. 46.)

[28] Revd Edward Frederick Holwell Dunnicliff, MA. Born Nottingham, 1901. Ed. Nottingham High School, Worcester College, Oxford and St. Stephen's House, Oxford. Curate of Blidworth, Notts. 1924-7. Vicar of Ollerton 1928-34. Vicar of St. Lawrence, Mansfield, 1934. (*Who's Who in Nottinghamshire*, p. 46.)

[29] *Dukeries Advertiser*, 20 Apr. 1934.

[30] See below, ch. 9 for the companies' attitude to sport.

[31] The Revd Percy Leeds.

[32] Interview, Charles Stringer, Harworth.

Ultimately, the reverend Mr. Comer was nominated by the Bishop to undertake the duties, commencing on the 1st of April, 1932.[33]

One reason for this development may be that the vicar of Bilsthorpe had encouraged the formation of a branch of the official trade union, the NMA, in Bilsthorpe in 1929, and had been marked by the company as a subversive influence, as the following report from colliery manager L. T. Linley reveals:

In the month of May I dismissed a Winding Engineman, named Edwards, for having violently pushed me out of the Winding Enginehouse without any provocation. The occurrence was brought about as a result of a visit I paid to the colliery on the eve of the 1st of May, when I found that the Power House attendant had left the house for at least 20 minutes, and, when he came in, he came from the Winding Engine House. I went to the Winding Engineman, Edwards, in quite a civil manner, and enquired why, as an experienced man, he had not told this man to return to his duties, and pointed out the danger of him being absent so long. He immediately turned on me and said he was not going to be bullied, and that he would turn me out of the Engine House, which he put into force.

This man Edwards has since done all that he possibly could to stop the Colliery, but the point is that he is a Church Warden and is being influenced by the Rector. He has done all he possibly could to get the men to stop work, and in the end, through the Rector's assistance, he was the means of a meeting being called in the Rectory field, to form a branch of the old Nottinghamshire Miners' Union, of which they have made him Secretary.

The Rector, in addition to finding the field, was at the Meeting, and responded to a vote of thanks, when he said that he was pleased they were forming a branch of the old Nottinghamshire Miners' Union (there is also a Spencer Union here) and that he should be pleased to lend them his Parish Room for Meetings. This is a clear case that the Rector is doing all he possibly can to cause agitation, and to create a bad feeling between the workmen, the Company and the Officials.

I have written this at length as I thought it only right that you should know what a really dangerous man the Rector is. I have no further suggestions to make than those which are already in hand as far as dealing with the Rector is concerned.

(Signed) L. T. LINLEY[34]

Often the employers merely approved proposals by groups of followers of various denominations to instigate religious

[33] A. E. G. Harmon, private history of the Stanton Company, typescript, 1960, vol. 5, p. 129.

[34] Stanton Company Colliery Committee, Appendix to Minute 559, 12 Aug. 1929.

activity in the villages. The history of the Baptist Church in Ollerton is a case in point. The impetus here came originally from outside the village. The Churches of the East Midland Baptist Association announced that:

by a resolution passed at our Assembly in Nottingham on June 2nd, 1926, the effort to establish work at Ollerton, and to build at once a School Chapel was commended to the Churches of the Association as the aim on which our united effort and mutual helpfulness should be concentrated for the year 1926–7.[35]

Such concentration was necessary because of the development of the North Notts. coalfield; Ollerton was becoming a town of considerable importance: 'Many Baptists are migrating to Ollerton from other districts, and they appeal to us for help. They are themselves a people who have a mind to work.'[36] The first meeting for those trying to form a Baptist Church in Ollerton was held at the Wesleyan School Room on 10 July 1926. It was agreed that the Revd A. Nightingale of Highfield House, Sutton-in-Ashfield, should act as a pastor for Ollerton.[37] One priority was to establish buildings for teaching and worship. There had been a tiny Baptist chapel at the contiguous village of Boughton which dated back over 100 years. The building had long since been derelict, and with its sale and other money available, the East Midland Association bought a piece of land in the business centre of New Ollerton for the construction of a school and chapel on the same site. The venture was 'earnestly commended to all who are concerned to see our Baptist witness adequately presented, and a spiritual home provided for our migrating Baptists and their families.'[38] 400 handbills were printed on 5 August 1926 for the advertisement of the Baptist activity in New Ollerton colliery village.

On 28 October of that year, the Revd Nightingale approached J. Bircumshaw, the deputy collieries agent of the Butterley Company, for help. The Baptists 'very respectfully solicit any help the company may deem fit to give. We are working in conjunction with the Company for the wellbeing of the Miners in the new Colliery Village: our part of the

[35] Butterley S6 311. [36] Ibid.
[37] Notts. CRO BP 19/1 Ollerton Baptist Church Minute Book, 1926–60.
[38] Butterley S6 311.

work being more on the moral and spiritual side.'[39] Bircumshaw was inclined to favour the request. On 30 October he wrote to his superior H. E. Mitton that 'I know these people are working very hard in connection with the moral and spiritual welfare of the people at Ollerton, and they deserve every encouragement in this work.'[40] On 24 November Mitton announced a donation of £50 by the Butterley Company towards the Baptist effort in Ollerton, and on 4 December 1926 the chapel, which seated 150, was opened by Mrs Bircumshaw.

The pattern of the Butterley Company encouraging but not instituting local religious bodies may be seen in connection with other institutions. The Ollerton Primitive Methodists owned a small chapel in the old village, but as its capacity was only fifteen, they asked Butterley in 1923 for a piece of land on the site of the new village of Ollerton: 'We are presuming that there may be a large influx of Primitive and other Methodists in the future and desire to cater for their requirements.'[41] Butterley decided not to grant any land, but did donate £50 in December 1927 towards the cost of a new Sunday School and Chapel. The construction of these buildings, which eventually cost £2,000, was started at a stone-laying ceremony on 18 February 1928 and opened on 2 June in the same year. The Primitive Methodist Home Missionary Committee provided £1,000, but the local followers had to find the rest. By 17 March 1931, the local minister, Clarence Pickering, was still attempting to pay off a debt of £861 by an Annual Effort. Butterley gave £5.

Alfred Thomas Narraway, the South Yorkshire Divisional Commander of the Salvation Army, wrote to the Manager of Ollerton Colliery on 18 August 1925:

> Seeing there is a possibility of the district around your colliery growing to some large dimensions, I am wondering whether it would be possible for the company to grant us a piece of land, and assist us with regard to the erection of our Building.
>
> There is hardly need for me to mention our work. Practically everybody knows we are out to help make men—better men—and also to be of help to the needy irrespective of class and creed.[42]

[39] Ibid., S6 311. [40] Ibid.
[41] Ibid., S6 311. [42] Ibid., S6 311.

H. E. Mitton agreed that Butterley should donate £50, and undertook to sell one rood of land so that a Mission Room could be erected. But he warned that 'it is desirable that it should not get into the press', and when Bircumshaw met Narraway on 14 December 1925 he reiterated that it was not to be advertised that the company was prepared to make such a grant, and no account was to be published.

It has always been recognized by historians that religion can be a powerful force either for political and social control or for disruption. It need not surprise us that the coal companies of the new Nottinghamshire field attempted to encourage regular church-going—or at least to make sure that any message that was being preached did not conflict with their own interests. After all, at an earlier stage of industrialization in the eighteenth and nineteenth centuries, factory owners had laid great stress on the religious guidance of their workmen. Sometimes it might seem that they too, like the Nottinghamshire coal owners who sponsored many denominations, were concerned less about which god was worshipped than that the virtue of respect was inculcated in the workmen and their families.[43] All forms of the Christian religion tend to avoid social attitudes which include notions of class interest and class conflict, the struggle between employer and employee.[44] To some extent socialism and Methodism, for example, offer mutually exclusive views of the fundamental principles of human life and society, as Bernard Taylor of Mansfield Woodhouse discovered. A Wesleyan lay preacher at the age of 15, between 1918 and 1921 he was 'converted' to socialism, finding his membership of the local Methodist chapel incompatible with his new beliefs.[45]

However, nor should we be surprised to find evidence of favour being shown towards one particular denomination. It seems clear that some men did find it advisable to adopt Catholicism in Ollerton. The personal beliefs of men who held power in the community did matter, just as they mattered in nineteenth-century rural England, where the allegiance of a

[43] Sidney Pollard, 'Factory Discipline in the Industrial Revolution', *Economic History Review*, 2nd ser. 16 (1963-4), 270.

[44] R. S. Moore, *Pitmen, Preachers and Politics* (Cambridge, 1974), p. 26.

[45] Lord Taylor, *Uphill All the Way* (London, 1972), ch. 4.

squire could determine the success or otherwise of the Methodist movement in a particular parish.[46] The same phenomenon could be found in industrial towns in the Victorian period. At Crewe, the brother of the Chief Mechanical Engineer changed his allegiance from chapel to church, and was followed by a number of lesser company officials and finally by many employees.[47]

As far as mining is concerned, it was alleged in the early years of the twentieth century that, in the Deerness Valley district of Durham, one coal-owning company, Pease and Partners, deliberately favoured Methodists in their recruitment of workers, and this accounted for there being a higher proportion of Methodists in some colliery villages than in others.[48] In the Dukeries, the colliery manager's movements on a Sunday were shadowed by many miners and their families.

The policy of the coal companies as far as religion is concerned is one more illustration of their energy and vision in attempting to maintain order in the pit and in the village. But to some extent, as with many of their other policies, it could be seen as anachronistic. By the 1920s, it was not easy to convert workmen to an ethic which would positively further the employer's aims. Religion was no longer such an influential force in mining communities, or in the national way of life, as it had been in previous centuries. The companies' success in their religious policy was limited, and it was negative. They could persuade more miners to attend church than perhaps otherwise would have done. They could silence subversive views from the pulpit in faviour of sermons which stressed the values of co-operation and sport. But whether religion played a great role in imposing the companies' morality on the miners is doubtful. By the 1920s, religious activity was but one of many features in a miner's life; and control over the form of religion in a community no longer had the power that it would have done in ages past. It was

[46] J. Obelkevich, *Religion and Rural Society: South Lindsey 1825-75* (Oxford, 1976), p. 35.

[47] W. H. Chaloner, *The Social and Economic Development of Crewe 1780-1923* (Manchester, 1950), pp. 153-4; Patrick Joyce, *Work, Society and Politics* (Brighton, 1980), pp. 176-7.

[48] Moore, *Pitmen*, p. 70.

true that the mining villages of the Dukeries coalfield still exhibited some of the features of the principle of *cuius regio, eius religio*—the Catholic converts of Ollerton attest to that. But this was not of itself sufficient to ensure peace and quiet. In order to achieve that, the companies would have to try to impose their influence in other spheres and in other ways.

Table 34. *New churches in the Dukeries colliery villages (dates of consecration)*

	C of E	Wesley	Prim.	Baptist	RC
Ollerton	1932	1927	1928	1926	1931
Bilsthorpe	1932	1933			
Blidworth		1932			
Langold	1928	1928			
Clipstone	1928	1932			
Harworth	1924	1925			

9. Leisure Time

'This was far better than hanging and slouching around street corners arguing about Socialism and Bolshevism'[1]

It is well worth enquiring how the miners and their families spent their leisure time in the early years of the new Dukeries coalfield. Not only can oral evidence offer a great deal of valuable information about this relatively rarely considered aspect of human experience, but it should be remembered that miners, like many working people, have failed to attach the same central importance in life to 'work' as opposed to 'play' that intellectual theorists have.[2] Also, leisure cannot be sharply differentiated from 'serious' matters such as the power structure of the mining community itself and the role of the miners and their employers within it. Caillois claimed that a society could be studied through its patterns of leisure, which reflect its values and aspirations, preferences, and weaknesses.[3] Given the powerful control that the Dukeries coal companies could exhibit over the lives of their employees, outside the pit as well as within it,[4] one might expect to find that here too, as in the case of paternalist mid-Victorian factory owners,

> play was not to be allowed any form of special licence; rather it had to be firmly and unequivocally integrated with the rest of life and securely anchored in orthodox morality . . . recreation grew to be accepted as a necessary amenity, a basic overhead in the maintenance of an industrial society.[5]

As in the case of the provision of educational services,[6] as in the case of the provision of basic amenities such as metalled roads, sewerage systems, and water schemes, the teething

[1] Henry Cropper, Mayor of Chesterfield, *Derbyshire Times*, 14 Jan. 1922.
[2] Alasdair Clayre, *Work and Play* (London, 1974), p. 122.
[3] Roger Caillois, *Man, Play and Games* (Paris, 1957; London, 1962).
[4] See above, ch. 4.
[5] Peter Bailey, *Leisure and Class in Victorian England* (London, 1978), p. 102.
[6] See above, ch. 7.

troubles of a new coalfield and of the new communities in it led to a considerable disruption in the leisure activities of the first miners to arrive in the Dukeries. Formal facilities were slow to appear, despite the concern of the colliery companies, who thought sport and recreation worthy of inclusion in their annual progress reports. At Thoresby Colliery, where sinking began in Spring 1925, a temporary Miners' Institute was not opened until 17 May 1929.[7] Bowling and putting greens were opened by the Bolsover Company colliery agent T. E. B. Young in 1932, but it was reported then that a sports ground was urgently needed as there was no decent playing field. In that year, Boys and Girls Brigades 'have not yet been formed but no doubt will be now a Village Hall is in course of erection'.[8] In fact the Brigades were founded on 25 October 1933 under the leadership of Mr Wyness, the colliery engineer, and they participated in their first annual camps at Rhyl (Boys) and Skegness (Girls) in Summer 1934.[9] The Thoresby Colliery Sports Ground and Pavilion were not opened until 16 May 1936.[10]

The story of delay was similar at other colliery villages in the new field. The Sports Ground at Clipstone, where sinking had commenced in 1920 and productive status had been attained in 1922, was still not in operation by 1937.[11] Other facilities, such as cinemas, also did not arrive until well after the mining population had settled in the Dukeries villages.[12]

Since the facilities for institutionalized sport and other leisure activities were not readily available to miners and their families in the early years of the colliery villages in the late 1920s and early 1930s, there was plenty of opportunity for less formal types of recreation. It may well be for this reason that Ollerton became well known as a centre for card-games, dominoes, and other indoor games which could be played with the minimum of equipment. This struck a journalist who wrote an article about 'A Coal Miner's Life Above Ground' there in 1932. Charles Graves reported that whist was the most popular card-game; very few played bridge,

[7] Thoresby Colliery Manager's Annual Reports, 1929.
[8] Ibid., 1932. [9] Ibid., 1934. [10] Ibid., 1936.
[11] Bolsover Company Managing Director's Annual Reports, 1937.
[12] For dates of licence of cinemas, see Table 35 below.

which has always been more popular amongst the middle classes. Money passed hands over solo, cribbage, and 'all fours', a select Irish game introduced by the sinkers. Dominoes was also played very frequently, and kibbitz cheating, in which a bystander revealed what was in a player's hand, was apparently an accepted custom. Rummy, pontoon, and tippit, a game in which a button was concealed, were pursued rather less seriously. Billiards and snooker flourished at Ollerton during this period, 'largely because a former Scottish amateur international lives in the village and gives lessons'.[13] This was Walter Donaldson, later world professional champion. The practice of friendly cheating seems to have extended to card-playing as well as dominoes:

> Indeed one or two of the card players were sort of acknowledged cheats. If you played with them you quite expected to be—they were very gentlemanly about it, though, if they had cheated you, would say 'you didn't notice that, did you?' I myself have been dealt a card off the bottom of the pack after being showed it, shown to be on the bottom of the pack, it was dealt to me in a pontoon hand, and I never saw it come from the bottom![14]

Another form of leisure which was cheap and readily available was to walk or cycle in the attractive Sherwood Forest countryside in which the colliery villages were situated. To some extent this was making a virtue out of a necessity:

> It was both strong for normal activities in walking, cycling, outdoor pursuits, because the colliery lies in a very pleasant wooded area—there are very nice walks and so on about—and a lot of them enforced by the poor amount of work available.[15]

Most of the children of the colliery villages knew every inch of the forest, and the estate parks of the neighbourhood.[16]

There were disadvantages in the rural nature of the coalfield, however. Remote and isolated, its inhabitants could

[13] Charles Graves, 'A Coal Miner's Life Above Ground', *The Sphere*, 23 Apr. 1932, p. 141.

[14] Interview, A. E. Corke, Ollerton.

[15] *Idem.*

[16] Interview, Harry Parnell, Edwinstowe.

afford to travel to town (Mansfield) only perhaps once a week, to sample its wider range of entertainment:

RJW What would you find in Mansfield that you couldn't find in Ollerton?

CB Well, entertainment mainly. Obviously there was a better selection of picture houses. I mean there wasn't a picture house in Ollerton.[17] And then also there was the old Queen's Variety on Belvedere Street at Mansfield, live music hall type of show. I don't think that was particularly well patronized, but I know that myself and my friends used to go to Mansfield on Friday night, just to go to that.

RJW So you'd say that the entertainment in Ollerton wasn't really sufficient?

CB No, I don't think it was, no.[18]

Shopping normally took place in the colliery village itself, either at a company store, where one was provided, or at the Mansfield or Worksop Co-ops, from travelling salesmen, or at the shops of the many small traders who came to the Dukeries along with the miners and their families. 'Tick' was almost universally available. On special occasions, and very rarely more often than once a week, shopping expeditions to Mansfield or Worksop took place. These were normally concerned with such items as footwear and clothing which could not be purchased in the village itself.

Cycling was all too familiar to the miners, for many of them rode several miles to work each day. Cars, however, were almost unknown, a fact which increased the sense of isolation. When asked how many miners owned cars in the pre-war period, one respondent replied: 'I think there were about three. My uncle had one before the war, black box Ford with yellow wheels, and he thought he was king of the road.'[19]

When more organized leisure pursuits did arrive in the colliery villages in the 1930s, the impression was very much that it was by courtesy of the colliery companies, who maintained a strong interest in all activities which took place in the communities, whether 'in company time' or not. All

[17] Until 1928.
[18] Interview, Cyril Buxton, Forest Town (Ollerton).
[19] Interview, Hilda Tagg, Rainworth.

the Dukeries coal companies vigorously encouraged sport, and went so far as to employ expert sportsmen in their pits in order to strengthen their representative teams:

They got a job if they could play cricket, if they could play football, if they'd got a good voice or if they were musical, because you see we'd got colliery bands, you know Ollerton had its own colliery band and Edwinstowe-Thoresby had its own later, and I think if they were any good at any of those sorts of things they got jobs and they kept them as well.[20]

This policy could illustrate the way the widespread regional origins of the miners in the new villages could influence the type of game that was played. At Harworth, the 'night-gaffer', Walter Tappin, and the undermanager, Harry Pedley, were both interested in bowls. They were able to induce good bowlers, especially from Lancashire, to come and work at the pit. Since these men knew only crown green bowls, and not flat green, Harworth has always looked to Yorkshire and the North for competition. Almost the whole of the rest of Nottinghamshire plays flat green bowls, favoured in the South and Midlands. It was a long-standing joke in Harworth that this pit, which was situated in Nottinghamshire, won the Yorkshire Cup with an all-Lancashire team![21]

The reason for the companies' sponsorship of sporting 'stars' was both to encourage pride in the harmony and unity of the pit, men and management alike, and to keep men out of trouble by offering facilities for 'healthy' sporting activity. The colliery sports grounds were seen as part of the general investment in the new coalfield. The companies held a monopoly of sporting facilities in the colliery villages, and the local teams were decidedly colliery rather than village teams. Often they reached a very high standard. Ollerton Colliery won the Butterley Company football shield in 1933/4, 1934/5, 1935/6, and were runners-up in 1932/3 and 1936/7, a dominance which led the Ollerton Colliery FC Secretary to suggest on 1 June 1938 that they should withdraw from the competition: 'Ollerton's football prowess has had the effect of making other teams reluctant to compete, or if they do compete, reluctant to meet Ollerton if drawn against

[20] Interview, Barbara Buxton, Forest Town (Edwinstowe).
[21] Interview, Neville Hawkins, Harworth.

them.'[22] Cricket matches were played as far away as Durham, which made it necessary to grant paid leave to the players.[23] Monty Wright was himself Chairman of the colliery Sports Club and Cricket Club.[24] There is an interesting parallel here with the dominance of the Savile Estate in the sporting and cultural associations of the 'old village' of Ollerton. In September 1934 Ollerton Cricket Club celebrated '50 years as a playing member of their popular captain, John Baker', the Savile agent, whose association with the club had begun in May 1884.[25]

The local newspaper items concerning Ollerton in these years of the private coal companies were heavily weighted in favour of sport. A cutting from September 1945 reads:

> I am told by Joe Perkins (well known in Nottingham as an old boxing referee), steward of the Ollerton Colliery Institute, that the management of Ollerton Colliery do what they can to make Ollerton a happy village and try to keep the people interested, which in the long run shows an increased coal output. Ollerton is like an oasis, miles from anywhere, yet they can boast of a team in the Midland League, and a good team too; a cricket team in the Bassetlaw League (champions in 1939); a reserve football team in the Notts and Derby Senior League. They have a male voice choir, silver prize band and even a tug of war team, which, by the way, was beaten in the final by Coventry Police at Solihull near Birmingham last Saturday. Mr M. F. M. Wright of Ollerton Hall renders every assistance he can to promote sport, and he thinks there is nowhere like Ollerton. Mr S. Thorneycroft, manager of Ollerton colliery, is also Chairman of the Football Club and Mr G. Belfitt (Butterley Co. Mining Agent) is also another great stalwart of Ollerton.[26]

In his Memoirs, Monty Wright himself wrote:

> I always felt that in a community like Ollerton that it was necessary to have good games teams and we had to promote good feelings amongst the chaps by having good teams. I played cricket myself and was captain of the Ollerton Colliery Cricket XI. I made certain that it was going to be a good side.[27]

There can be no doubt that sport was seen as one more

[22] Butterley W2 440. [23] Ibid.
[24] *Newark Advertiser*, 11 Jan. 1928.
[25] *Dukeries Advertiser*, 21 Sept. 1934.
[26] Butterley P9 292 'Propaganda' file.
[27] M. F. M. Wright, Memoirs, p. 40.

legitimate field of activity on the part of the colliery companies' management:

RJW Why do you think it was so important for the colliery company to encourage, to have a good sports team and a good band?

CB Well, I have definite ideas about that. If their employees were sort of happy in their spare time likely they'd work better and they'd not have union meetings complaining about things, they'd be happy playing their cricket, or their football, or their music; in other words it was good for them, it kept the troops happy, so to speak.[28]

This was by no means an unusual view in the inter-war Midlands coalfield. In 1922 Henry Cropper, the Mayor of Chesterfield who was later to defect from the Labour party to become a Liberal parliamentary candidate, told a Chesterfield football club that: 'One of the reasons why this country would never witness a political or social revolution or upheaval was because the average Englishman is immersed in sport . . . this was far better than hanging and slouching round street corners arguing about Socialism and Bolshevism.'[29] D. N. Turner of the Staveley Company, whose subsidiaries sunk both Blidworth and Firbeck Main pits in Nottinghamshire in the 1920s, expressed similar sentiments: 'Cricket taught people more than just how to use a bat and ball. It taught them to play cricket through life, in everything they did. Generally the best sportsman was the best workman, and the most useful citizen in the end.'[30]

Sport, therefore, was encouraged by the Dukeries coal companies because of its positive moral effect as well as because it could act as a harmless safety valve to divert men's spare time and passions from more controversial matters. The employers spent a great deal of money on the organization of sport, but it was not entirely out of their own pockets. Contributions towards the sports fund were usually included among the semi-compulsory 'stoppages' at the wages office. Similarly the colliery bands were regarded as of great importance. As the mining agent told the Ollerton colliery manager

[28] Interview, Cyril Buxton, Forest Town (Ollerton).

[29] *Derbyshire Times*, 14 Jan. 1922, quoted in J. E. Williams, *The Derbyshire Miners* (London, 1962), p. 790.

[30] *Derbyshire Times*, 29 Nov. 1935.

in 1937, the players of the vital instruments in the band were not to be given notice of dismissal, 'even if it means sacrificing a less important employee'.[31] There were no miners-union or lodge bands as were found in older British coalfields such as Durham.

It is clear that bandsmen, like talented sportsmen, were regarded as especially valuable for boosting the colliery's prestige. The local newspapers were keen to report the success of the colliery bands in competition, such as that of the Harworth silver prize band in 1931.[32] It may well be that in Nottinghamshire in the 1930s, just as for James Hudson in the 1850s, 'the manufacturer finds it PROFITABLE to form schools and factory libraries, to rear amateur bands of musicians among the workmen'.[33] Perhaps the idea of rational recreation, of a play discipline which might reform popular leisure pursuits to back up industrial work discipline was not confined to mid-Victorian factory masters. It would certainly not be possible to write, as the authors of *Coal is Our Life* did in the case of post-war Featherstone, that 'the colliery itself is only of slight significance in leisure activities'.[34] It cannot be denied that whatever intentions of 'social control' the companies may have had in providing the colliery bands, they offered a great deal of pleasure to the inhabitants of the Dukeries coalfield, just as the sports encouraged by the companies did. But the fact remains that the provision of opportunities of this nature was overwhelmingly in the hands of a single authority.

The colliery company officials played a leading part in the voluntary clubs and associations to be found in all the mining villages in the new Notts. field:

> You see, under private enterprise, the colliery company, the manager, was more or less the figurehead in the village, he and his wife were almost like the country squire and his wife . . . and then you went down to the undermanager at the colliery, and the chief cashier at the colliery, or the Chief Clerk as you called him, he was the concert secretary, the club secretary, all the little things that were going on in

[31] Butterley W2 440.
[32] *Worksop Guardian*, 21 Aug. 1931.
[33] Peter Bailey, *Leisure and Class in Victorian England* (London, 1978), p. 44.
[34] N. Dennis, E. Henriques, and C. Slaughter, *Coal is Our Life* (London, 1956), p. 119.

the village, he was the secretary for all of those things, he had a finger in every pie, sort of thing![35]

Similar personnel were to be found guiding the popular associations for the younger members of the communities, the Boys and Girls Brigades, whose activities regularly figure in the colliery managers' monthly and annual reports to company boards.[36] The Brigades clearly offered welcome opportunities for the children of the villages to get away to camp for a week each year:

RJW Were they encouraged by the company?
FT Yes. They provided camps. There was a camp at Rhyl, a camp at Skegness, I think they went for a week at a time.
GC And Scarborough, they used to go to Scarborough.
FT Yes, and I think they'd manage about three weekends and a whole week at one of these camps for all the younger element, yes, a good thing, as I say the place itself was absolutely the tops in the industry, there wasn't any place like it—
DT It was marvellous, at Skegness, which is now the Derbyshire Miners' Home. They used to come from all the collieries and all go together, march up there with the different bands from all the different collieries, and the Boys Brigades' bands, and when they played 'Georgia', it was out of this world, it was lovely, they've nothing like it now.[37]

This one week's release from the routine of life in the colliery villages was very much due to the company's sponsorship, in these years before a family holiday could be taken away from home on a regular basis.

St. John's Ambulance Brigades and local Nursing Associations were also regarded as company operations which were mentioned in annual colliery reports to the board. Ambulance classes were strongly encouraged to the point of being semi-compulsory for underground colliers in many of the villages. This was a sensible policy, given that each year there might be as many as 1,000 accidents requiring first aid treatment, regardless of minor injuries which went unreported, in each of the pits throughout the inter-war period.[38] Often the

[35] Interview, Barbara Buxton, Forest Town (Edwinstowe).
[36] See, for example, the reference to Mr Wyness, p. 190. above.
[37] Interview, Frank Tyndall, George Cocker, Dorothy Tyndall, Edwinstowe.
[38] Thoresby Colliery Annual Reports, Accidents and Safety: the smallest number of accidents reported in any year between 1930 and 1946 was 630 in 1938; the most 1,407 in 1943.

Ambulance Brigades were treated to a week's holiday at the seaside by the company. The colliery manager was normally the president of the local brigade:

He did make everybody go to ambulance classes, everybody. And then they had the parade every month, you know, used to parade around the village, probably go to Bawtry some times, it would just depend on what were going on. And Wright,[39] he didn't know the first thing about first aid, and he used to walk in front with all those medals on and stripes, he'd never been to an ambulance class in his life! But it didn't matter, he was it.[40]

Besides organizing the Boys and Girls Brigades' annual camps, most of the colliery companies also put on summer day-trips to the seaside, usually the Lincolnshire coast, for the whole population of their colliery villages. These proved immensely popular, especially with the children:

But I can remember when we were children, it was the day of the year for a lot of us, because we never went away when we were children, my father and mum couldn't afford to take us, and it was the day of the year because you all had your best clothes on that day and your mum was up early packing all the sandwiches up. We used to have, I think, from Edwinstowe, it was four or five trains left the village on that one Sunday of the year, and there were all the children, you used to have labels pinned to your front with your name and address and your age on, and we used to have free rides on the roundabouts, and we were given, I think your mum was given so much for your food, I can't remember how much it was now.[41]

This subjective evidence is corroborated by the reports of the colliery day-trips in the Bolsover Company records. In 1938, for example, 1,534 people were taken to Skegness on 18 June in three special trains, and 612 went from Clipstone to Scarborough in the same year.[42] The Butterley Company archives record an even more ambitious plan, to visit London in June or July 1938, which was cancelled as it became clear that it would make a substantial loss.[43] Sometimes outings were organized by the company for more specific reasons.

[39] William Wright, manager of Harworth Colliery.

[40] Interview, Tommy Jenkins, Harworth.

[41] Interview, Barbara Buxton, Forest Town (Edwinstowe).

[42] Thoresby Colliery Manager's Annual Reports, 1938; Bolsover Company Managing Director's Annual Report, 1938.

[43] Butterley W2 440 'Butterley Co. Outing 1937–40'.

When the colliery manager and resident of Ollerton Hall, Montagu Wright, was married in June 1934, an opportunity was taken to impress the local population with the benevolence and grandeur of the company's management. A special train was hired which took hundreds of villagers to London for the wedding. The local paper's headline was 'Over 300 Ollerton People Attend Fashionable London Ceremony'.[44] The Abbot of Ampleforth officiated. One is reminded of the way that the wedding of a resident paternal squire in a nineteenth-century agricultural district would be translated into a communal celebration, as if 'the squire and his relations were the parochial version of a royal family'.[45] In nineteenth-century industrial Lancashire too, the domestic landmarks of the employers' families did not go unnoticed. Births and marriages were feted, the coming of age of sons was celebrated, the dynasty was rejuvenated and renewed.[46] In the Midlothian coalfield in 1867, the birth of a son and heir to the paternalist coal-master Robert Dundas was celebrated by a feast for 400 of his miners.[47]

The coal companies' influence extended too into leisure activities which were less directly their responsibility. As seen above, in several of the colliery villages it was against the terms of the lease of company houses for miners to keep dogs, especially whippets, or other pets. Gardening was an activity frequently enforced by the employers. Of course, many miners relished the opportunity both to indulge in outdoor pleasures far removed from the atmosphere of the pit and to grow vegetables to help feed the family. But the element of compulsion remained: 'Well, of course, you cultivated your garden, partially as the Butterley Company liked you to do your garden, firstly, and secondly as a means of feeding your family.'[48] There were many expert gardeners in the villages, and there were active and competitive horticultural associations; the Bilsthorpe Gala of 1933 was organized by the Garden Holders Association.[49] But the companies

[44] *Dukeries Advertiser*, 29 June 1934. [45] Obelkevich, p. 36.

[46] Joyce, *Work Society and Politics* (Brighton, 1980), p. 183.

[47] John A. Hassan, 'The landed estate, paternalism and the coal industry in Midlothian, 1800-1880', *Scottish Historical Review*, 59 (1980), 86.

[48] Interview, J. E. Smith, Ollerton.

[49] *Newark Advertiser*, 22 Aug. 1933.

undoubtedly encouraged well-kept gardens as part of their campaign to maintain the order and neatness of their 'model' villages as a symbol of the 'due order and degree' which they liked to see in these communities' social relationships.[50] One is reminded of the epitome of a paternalist employer, Trafford of Wodgate in Disraeli's *Sybil*:

> When the workpeople of Mr. Trafford left his factory, they were not forgotten. Deeply had he pondered on the influence of the employer on the health and content of his workpeople. He knew well that the domestic virtues are dependent on the existence of a home, and one of his first efforts had been to build a village where every family might be well lodged . . . in the midst of the village, surrounded by beautiful gardens, which gave an impulse to the horticulture of the community, was the house of Trafford himself, who comprehended his position too well to withdraw himself with vulgar exclusiveness from his real dependants, but recognised the baronial principle, reviving in a new form, and adapted to the softer manners and more ingenious circumstances of the times.[51]

It is notable that however keen miners may have been on gardening for its own sake, residents of the Dukeries colliery villages have frequently observed that one of the most noticeable effects of nationalization was that many miners immediately ceased to bother about keeping up the appearance of their gardens. Zweig claims that only 10-20 per cent of miners in the nationalized industry were active gardeners.[52]

The companies even managed to become involved in that most common of miners' pastimes, drinking. Licensed premises in the colliery villages were generally confined to the Miners' Institutes, owned, built, and run by the employer. There was a pub in Bilsthorpe in the early years of that village; the Stanton Arms Hotel, owned and built by the Stanton Coal Company, which remained the only public house in the village till its sale to the Samuel Smith Old Brewery of Tadcaster in March 1946, towards the end of the era of private ownership of the mines.[53] In Ollerton there

[50] See below, ch. 12.

[51] B. Disraeli, *Sybil* (1845), Bradenham Edition, pp. 211-12. The managers of the new Dukeries pits also lived in the villages.

[52] F. Zweig, *Men in the Pits* (London, 1948), p. 106.

[53] A. E. G. Harmon, privately circulated history of the Stanton Company (7 vols., 1960), vol. 5, p. 105.

were coaching inns in the 'old village', but never a pub in the colliery village:

RJW How many pubs were there in Ollerton at that time?
CHG About that time there was the Royal Oak, the White Hart and the Hop Pole.[54] They wasn't allowed through some ruling of the Butterley Company—oh, and I'm forgetting the big pub, the Plough, that were open in '25, that were open when the pit started sinking—but there were some ruling, I don't know whether it came from the Butterley Company or from the estates, that they couldn't build a pub in Ollerton, and there still hasn't been one built.[55]

In February 1928 the Brewster Sessions of Worksop rejected the application of J. E. W. Wilson of Droversdale Road, Bircotes, to open an off-licence in the colliery village for the miners of Harworth Colliery, despite the fact that the centre of the colliery village was 1,360 yards distant from the Galway Arms, an inn in the old village of Harworth.[56] The result of the shortage of pubs in the colliery villages was that even while drinking, the miners usually found themselves in a company institution. In the Harworth dispute of 1936/7, the company-owned Institute was the social centre and drinking place for the men who continued to work; the strikers congregated in the more independent Comrades Club. The Derbyshire policemen brought in to protect the 'blacklegs' were billeted in the upper rooms of the Institute.[57]

In the cinemas too, the companies often provided an essential ingredient: light and power supplied from the colliery. The Butterley Company was supplying electricity for the Ollerton Picture House at 2*d.* a unit in 1936.[58] Apparently, the colliery manager Monty Wright partly owned the Ollerton cinema; certainly the 35 mm film made of the day-trip to London to celebrate his wedding was shown there.[59]

[54] The pubs mentioned were all situated in the old village of Ollerton, and not in the colliery village. The Hop Pole was named after the hops used for brewing beer in Ollerton up to the nineteenth century. Hops still grow wild in the hedgerows and banks around Ollerton.

[55] Interview, C. H. Green, Ollerton.

[56] *Worksop Guardian*, 3 Feb. 1928.

[57] Interview, Stan Morris, Harworth.

[58] Butterley L1 226 'Ollerton Village 1934–7'.

[59] Correspondence, Stephen Wright, Ampleforth.

In August 1929 the Stanton Company decided to convert the Bilsthorpe Village Hall for use as a cinema on three nights each week; the equipment and the provision of seating accommodation for 408 persons cost the company £500. The first performance in the Village Hall took place on 12 December 1929, and for the first few months it was found that costs amounted to approximately £15 per week, including £6 to £8 for the hire of the films, while takings averaged £18 per week.[60] This example of the coal company operating the village cinema was unusual even in the new Dukeries coalfield, and in most of the villages commercial cinemas were established, as we can see from the Nottinghamshire County Council's Clerk's Department's files of the registration of cinema licenses, shown in Table 35.

Table 35. *Cinemas licensed in the Dukeries colliery villages*[a]

license no.	cinema	licensed from	to
5/2/46	Edwinstowe Hall	1 Jan. 1924	31 Jan. 1948
5/2/49	Picture House (Palace Cinema), Langold	1927	1959
5/2/50	Scala Cinema Theatre, Langold	1927	1937
5/2/52	The Picture House, Boughton Rd., Ollerton	14 Jan. 1928	present
5/2/58	The Cinema House, Scrooby Rd., Harworth	6 May 1930	Oct. 1960
5/2/71	The Scala Cinema, Mansfield Rd., Blidworth	24 Dec. 1935	30 Jan. 1960
5/2/76	The Major Cinema, Mansfield Rd., Edwinstowe	4 Nov. 1936	28 June 1958
5/2/78	The Ritz Cinema, Mansfield Rd., Clipstone	9 Jan. 1937	2 Feb. 1959
5/2/90	Scala Cinema (reopened), Blidworth	14 Feb. 1960	21 May 1962
5/2/91	Ritz Cinema (reopened), Clipstone	21 May 1962	3 Nov. 1962

[a] Notts. County Council Clerk's Department, Registration of Cinemas, Notts. CRO CC CL 2/5.

[60] Stanton Colliery Committee Minute 559, 12 Aug. 1929; Bilsthorpe Colliery Monthly Report, Nov. 1929; Harmon, Stanton history, vol. 5, p. 130.

A. E. Corke came to Ollerton to operate the cinema there in the 1920s:

RJW How often did people go to the cinema, did they go once a week or at a change of programme?

AEC It was certainly very well attended, and people would even complain that a certain popular film had been, and they couldn't get in to see it. I always remember popular children's matinees, every Saturday, and more during the holidays. Every Tuesday morning there was a miners' matinee, so that men on the back shifts could have gone and seen it if they'd wanted to. At the weekend, of course, when the shifts were changing, they could go to the ordinary evening shows.

RJW How often nightly were the films shown? Just once?

AEC Generally twice nightly. But during the period of depression in the village it did get down to once.

RJW And what kind of films were shown? Were there films from America?

AEC Oh, a very catholic taste. No, in films you could usually not get the very best of films unless you had also booked some supporting programmes of others, so you had to take the rough with the smooth. Certainly Westerns.

RJW Did these films come round when they first came out or did you feel you were a few months behind in Ollerton?

AEC Ah, you see there again the major cinemas having to pay more for their films could bar other films from showing within a certain radius, so they did not attract custom.[61]

The delay before major films reached the Dukeries, and also the wide variety of films shown at these cinemas in a typical year may be gleaned from a perusal of the advertised programmes in local newspapers. In May 1941, for example, we find Humphrey Bogart and Ann Sheridan in *It All Came True* (released in 1940) sharing the bill at the Major Cinema, Edwinstowe, with George Formby's comedy *It's in the Air* (1938).[62] In June the Western *Drums Along the Mohawk* (1939) was showing in Ollerton,[63] while in September *Philadelphia Story* (1940), a sophisticated comedy about the higher reaches of American society starring Cary Grant and James Stewart reached Edwinstowe.[64] By January 1942, broad comedy was again on the programme: Abbott and

[61] Interview, A. E. Corke, Ollerton.
[62] *Worksop Guardian*, 2 May 1941.
[63] Ibid., 20 June 1941.
[64] Ibid., 5 Sept. 1931.

Costello's *Rookies* (1940) was on in Harworth.[65] In August 1941, customers were being asked to pay 1*s*. 2*d*. for a balcony seat at Ollerton Picture House, and stall seats cost 8*d*., 10*d*., and 1*s*. 2*d*.[66]

If the colliery companies played an important role in the leisure activities of the Dukeries villages, leading citizens of the 'old' villages also dominated various voluntary associations. When a branch of the Junior Imperial and Constitutional League (the 'Junior Imps') was started at Edwinstowe in 1928, its inaugural committee consisted of Lady Titchfield, wife of the local MP; Lady Sibell Argles, daughter of Earl Manvers and wife of his agent, Hubert Argles, who handled his colliery business; Lady Eveline Maude of Cockglode House; and H. Simmonds, the Conservative agent for the Newark Division.[67] The leadership of the junior members of the Imps also devolved upon members from respectable backgrounds. In 1932 the Edwinstowe branch was dominated by the daughters of Percy Morley, local garage proprietor and hirer of automobiles to the Dukeries aristocracy,[68] and Southwell Rural District Councillor for Edwinstowe.[69] There was a Conservative Angling Club at Ollerton in November 1932,[70] and when a Poultry and Rabbit Club was founded at Edwinstowe in 1942, its first president was the manager of Thoresby Colliery, Charles Edward Woodward.[71]

Not all leisure time activity in the Dukeries coalfield was as 'respectable' as this. We must consider too what might be called 'underground' leisure—those activities which at least in theory were forbidden by the law of the land. The crime rate in the tightly-knit colliery villages was very low—often there was no local police constable but only a pit bobby.[72] Vandalism in the impeccably kept model villages was almost unknown. But one form of illegal activity was widely practised. Throughout the inter-war period betting shops were officially banned in England. Nevertheless there *were* betting shops in all the colliery villages before the Second World War:

There certainly were bookies' premises before the war and they were a

[65] Ibid., Jan 1942.
[66] Ibid., 22 Aug. 1941.
[67] Ibid., 29 Jan. 1928.
[68] See above, p. 134. for a description of Morley.
[69] *Newark Advertiser*, 27 Jan. 1932.
[70] Ibid., 30 Nov. 1932.
[71] *Worksop Guardian*, 26 June 1942.
[72] See above, pp. 98–9.

considerable function of the social scene. Normally more of a hutment, with a cinderpath floor, telephone installed, rough bench seating and a Racing Form paper, known as a 'Tishy', pinned to the fence-wall. From time to time an uniformed policeman was to be seen taking up a position adjacent to the bookmakers, in full view of the customers, and he was there to make 'observations' . . . this was the prelude to a raid on the premises by uniformed officers during which much of the business proceeded as usual. The later court proceedings and the imposed fines upon both the bookmaker and the devotees were regarded as normal overheads of the business. There were never any closure orders always provided that the conduct of affairs was in an orderly manner.[73]

Much the same was the practice in Harworth: 'They wasn't legal. In fact many times these was raided. You know, you'd get tipped off that they was coming, and the police would come in and raid them and the men would scatter in all directions. The bookies generally paid their fines for them.'[74]

More concrete evidence supports this oral testimony. The local press, for example, frequently reported raids by the police on the clandestine betting shops. In June 1934 Ernest Pearson of Wellow Road, Ollerton, was summoned to Worksop Police Court for keeping a betting house in respect of premises, and five men were summoned at the same time for resorting. The sum of £34 was seized in the raid; Pearson was fined £20 and the resorters were asked to pay the costs of the case. Pearson was also summoned for keeping premises in the neighbouring colliery village of Edwinstowe, and seven people for resorting there. The court's business was concluded with the summoning for premises of R. J. Maund of the Sherwood Café, Ollerton.[75] It is clear that betting shops were common in the mining villages, and that occasional raids were simply seen as part of the 'overheads' of the shop. In 1938 there were again raids at Ollerton and Edwinstowe, and 25 cases came before Worksop Police Court. That a wooden hut in Sherwood Drive, Ollerton had been used for ready money betting had been observed by a policeman (PC Clarke) on 2 and 3 August, and he had raided it in company with Sergeant Bird on 12 August. Ernest and Clarence Pearson were found guilty of keeping and using a betting house, and fined £20 and

[73] Correspondence, A. E. Corke, Ollerton.
[74] Interview, Stan Morris, Harworth.
[75] *Dukeries Advertiser*, 22 June 1934.

£5 respectively; both men had previous convictions (see above). These men had also kept a similar hut on East Lane, Edwinstowe. In these cases too, men were charged with resorting and asked to pay costs.[76]

Gambling was not restricted to betting on horse-races. At Langold, for instance, tossing rings was very popular:

> Well you see, the gambling laws were very strong in those days. There was two florins they used to use, or two half crowns. Two silvers, and they used to get something like three or four hundred men, and they'd pay watchers, there were watchers two or three hundred yards around the ring, and there were various little spots that were glades in the wood, and they'd vary 'em. In fact it's even been known for the village policeman to go and play, never mind catch them! Occasionally the police would raid.[77]

Another form of forbidden activity was that connected with illicit sexual behaviour. In general the mining villages exhibited a strong sense of conventional morality. Prostitution seems to have been almost unknown: 'Upon reflection I cannot think of any evidence of easy virtued women in those times. This was then a smaller community so that "carryings-on" could not remain anonymous and the housing was of the Colliery Compound type, and as such very strictly controlled as to tenancies.'[78] Even 'living in sin' was discouraged: 'We had very few living together, marriage was marriage, there was no doubt about it.'[79] As late as 1944 the Ollerton company policeman 'Bobby' Healey reported to the Butterley management a case of the estranged wife of a cornet player in the band, a 'sound workman'. She was living with another Butterley employee. Healey made it quite clear that the sinful couple could not live in a company house in New Ollerton, a view complied with by his superiors.[80]

In many ways leisure activities in the Dukeries coalfield were typical of the whole of pre-Second-World-War England. It would be absurd to claim that social control was the sole function of the horticultural societies, the sports clubs and the other institutions through which the miners could enjoy

[76] *Ollerton, Edwinstowe and Bilsthorpe Times*, 26 Aug. 1938.
[77] Interview, Ernest Baker, Langold.
[78] Correspondence, A. E. Corke, Ollerton.
[79] Interview, Charles Stringer, Harworth.
[80] Butterley 09 224.

fresh air, relaxation, and safety as a contrast to their dangerous and arduous jobs. The Ambulance Brigades, the Boys Brigades, and the Nursing Associations fulfilled useful and non-controversial social functions no matter if the colliery management was in charge, as in a different way did the cinemas and dancehalls, where escape could temporarily be purchased from the routine of the colliery villages, where 'steam could be let off', and where marriage partners for life could be found.

Yet there is still a point in stressing how even in the sphere of leisure the prevailing aura of company domination, order, and competitiveness could be found. Inter-village rivalry was encouraged to an unusual extent. Skill in sport and music was valued for rather less generous reasons than usually found in pluralist societies. There can be no doubt that the coal companies considered that the form the recreation of their employees took was of concern to them, and bore indirectly on their investment in exploiting the rich mineral resources of the Dukeries: the evidence of the company and colliery reports testifies to that. The employer's omnipresence at the head of voluntary and recreational organizations, as with nineteenth-century factory paternalists, placed the miner and his family 'in the shadow of his name'.[81] The inequalities of the gift relationship, the omnipotence of the donor, were hammered home by the symbolic (and sometimes literal) presence of men like Monty Wright of Ollerton and William Wright of Harworth at the head of institutions such as the Ambulance Brigade and its parade. Community was nurtured, loyalties bolstered; familiarity and friendliness, along with respect, tempered subservience.[82] Leadership and influence was the other side of the coin of authority and coercion.[83] Such an interest shown by employers in every aspect of their employees' lives could be regarded as unhealthy; and it was certainly not to be the road on which Britain and its coalfields travelled after the Second World War.

[81] Joyce, *Work, Society, Politics*, p. 170.
[82] Joyce, p. 165.
[83] Ibid., p. 102.

10. War and Nationalization

The impact of war: 'The men have got the upper hand and they know it'[1]

The social history of the British coalfields was affected by the Second World War in a different way from that of the rest of the country. Due to the status of mining as a protected industry, the communities were not deprived of adult male 'breadwinners':

RJW And how did the war affect miners? Presumably they didn't go to fight, they stayed in Ollerton?

AEC Well, some did of course, some had already belonged to Territorial Associations, some would volunteer, a few did. Some won honours, certainly, and lost their lives, and prisoners of war that came back. But during the war, one barely noticed that these people were missing unless they were particular friends of yours.[2]

A substantial minority of Dukeries miners did join up before the restrictions imposed by the Essential Work Order of May 1941 came into force. In that year the manager of Rufford Colliery reported to the Bolsover Company's managing director that 115 of his men were in HM Forces. This was of particular annoyance to Bolsover because it meant that their tied company houses could not be occupied by a working miner.[3] By August 1941 the Butterley Company were actively trying to recover their ex-employees from the Forces.[4] This was, however, not always an easy task. Some men relished the opportunity to escape from the hard, routine, underground life to discover new places and new experiences in the Army:

MW They took the opportunity to get out of the coal mines and go into the Forces.

[1] Thoresby Colliery Manager's Annual Report, 1943.
[2] Interview, A. E. Corke, Ollerton.
[3] Bolsover Company Managing Director's Annual Report, 1941.
[4] Butterley H3 224.

RJW They welcomed that opportunity, did they?
MW Oh, they did, yes. Not everybody as I say wanted to be a miner anyway. My brother certainly had no intentions of ever being one.
RJW But when it became a reserved occupation there were various attempts to get people back into the mining industry.
MW Yes, there was, and of course they simply didn't get the volunteers at all, I mean nobody wanted to do a job like that unless they were absolutely forced to do.[5]

Nevertheless, despite what the incidence of war may tell us about the relative attractions of the lives of miners and soldiers, the fact remains that the majority of men did stay in the colliery villages. The Dukeries did of course go through the motions of preparing itself for war. Little 'active service' was seen even in this 'total war'. Stray bombs fell in Thoresby Park and on Thoresby woodyard[6] as a crippled plane jettisoned its load. Three bombs fell on the corner of Poplar Street in New Ollerton, but did not explode.[7] Only one serious and deliberate attack was ever made:

It came one night, Lord Haw-Haw had been shooting his mouth off on the wireless like that he was going to shut Thoresby up because at that time it was supposed to be the best and biggest pit in Europe, Thoresby Colliery. In fact it was the first all-electric pit: there was no pit chimney at Thoresby, isn't now. And he'd threatened to shut this pit up and he dropped a bomb two miles away, didn't affect the pit like at all, just blew a few trees up in the forest.[8]

In general, however, the air raid sirens only sounded as the German planes passed over on their way to Sheffield.[9] Otherwise, the most direct impact of war on the colliery villages lay in the numerous petty inconveniences and shortages that all who lived through the experience of war on the home front remember: the blackout and the other air raid precautions, the rations and the luxuries, the billeted soldiers and the drill. In the Dukeries, some of these minor disruptions are remembered still, revealing in their own way the routine thus broken. In December 1940, complaints were made about the way that the Butterley Company cut off the electricity supply

[5] Interview, Marjorie Wilson, Blidworth.
[6] Interview, Charles Tyndall, Edwinstowe.
[7] Interview, J. E. Smith, Ollerton.
[8] Interview, J. H. Guest, Edwinstowe.
[9] Interview, Barbara Buxton, Forest Town (Edwinstowe).

it provided for the colliery village of Ollerton at night, which ensured that the blackout regulations were not broken but which also brought home the company's control over their lives to the miners in a forcible, symbolic manner.[10] In July 1940, Butterley had themselves complained that soldiers billeted in the village in miners' houses were damaging the much vaunted lawns, on which the miners themselves were not allowed to step, by playing impromptu games of cricket on them. But unable to employ the threat of dismissal in this instance, the company could only protest to the captain of the relevant company at the Proteus Army Camp (which was situated near Ollerton throughout the war).[11] The residents of Ollerton also had some cause for regretting the arrival of the soldiers. At least one ex-miner associated the war with the resulting shortage of beer:

RJW What effect did the fact that we were at war have on the village?
CHG Well, one of the effects, of course it was the shortage of beer, and after Dunkirk they sent—a lot of the soldiers came back here and they were put in Thoresby Park. Well, it's still an army camp now, they call it Proteus Camp. And of course they came up to our pubs and clubs, and we didn't like them really coming and supping our beer.[12]

Shortage of beer was quite a serious matter as far as miners were concerned. Besides the high rate of dehydration caused by their work, the pubs and clubs were social centres for the men in the colliery villages. So despite their higher wartime wages,[13] a resident of Ollerton could say of the war:

I don't think they thought it much because prosperity is a matter of what your possessions can be, and certainly there wasn't even much beer then. Miners got in heavy industry, they got extra rations, which was one thing that was taken in consideration as it were, but then shops, as in other areas, had very little to offer, so increased savings didn't allow them much at all.[14]

Many other peacetime activities also had to be curtailed or suspended. Entertainment outside the village could not be sought so readily:

[10] Butterley H3 224.
[11] Butterley H3 224.
[12] Interview, C. H. Green, Ollerton.
[13] See below, pp. 214.
[14] Interview, A. E. Corke, Ollerton.

Well, you became more closely confined because your transport in and out of the village was cut bad. Your last bus from town[15] was half past eight. You see, if you went to a film in town, you very often had to come out before it finished, or you walked three miles, four miles home, that sort of thing. And of course you couldn't go away on holiday. Your holidays were spent in the village but you see the Mines Welfare Committee took that over and we had what they called the Holidays at Home. And it was a week of organized events.[16]

The *Worksop Guardian* described the Holidays at Home for August 1942 as including boxing tournaments, poultry and rabbit shows, bands, vegetable and flower displays, whippet racing, an athletics demonstration, and football matches.[17] The pre-war annual outings organized by the company, and attended by almost the whole village, had to stop. In 1940 the annual outing for Bolsover employees to Skegness was cancelled, together with the Boys and Girls Brigades Camps.[18] That year too, for the sake of war production the miners' paid holiday of a week was replaced by two weekends, 23 to 26 August and 14 to 17 September.[19] Due to the blackout and the alternative activities organized by the cadet corps, the Boys Brigade at Rainworth had diminished from a 1939 total of 115 to sixty-five by 1941.[20]

Besides these negative effects of the war on the everyday lives of the inhabitants of the colliery villages, there were certain positive preparations that had to be made. As early as 1938, trenches and observation posts were constructed as early Air Raid Precautions at Edwinstowe.[21] Air raid shelters were provided at Rufford Colliery in 1939.[22] By 1940 101 men had been trained in the ARP at Edwinstowe.[23] The Butterley Company acquired land for allotment purposes at Ollerton in 1939 in an attempt to improve wartime food

[15] Mansfield.

[16] Interview, Hilda Tagg, Rainworth. Rainworth was the nearest of the 1920s villages to Mansfield, so these remarks apply even more strongly to the other colliery villages.

[17] *Worksop Guardian (Ollerton, Edwinstowe and Bilsthorpe Times)*, 14 Aug. 1942.

[18] Thoresby Colliery Manager's Annual Report, 1940.

[19] Ibid.

[20] Bolsover Company, Rufford Colliery Manager's Annual Report, 1941.

[21] Thoresby Colliery Manager's Annual Report, 1938.

[22] Bolsover Company. Rufford Colliery Manager's Annual Report, 1939.

[23] Thoresby Colliery Manager's Annual Report, 1940.

production.[24] The RASC requisitioned Edwinstowe Village Hall in 1940.[25] In 1941 the Bolsover Company tried to introduce a War Savings scheme for the employees at Thoresby Colliery, but met initial resistance: 'It is very difficult to persuade the men to contribute. Incidentally it is causing a good deal of absenteeism and grumbling.'[26] The reason for this was that men suspected that they would see no more of their hard-won wages. Also at Thoresby, an air training corps was founded in July 1941. Established as Squadron no. 1491, by the end of the year it had a strength of eighty-eight. As an interesting reflection of the social structure of the village, the colliery manager was the chairman of the corps, and the headmaster of the Church School was its commandant.[27] At Ollerton, Butterley Company mining agent W. S. Fletcher presided at the annual Home Guards dinner in March 1941. The platoon commander was J. T. P. Foster, landlord of the White Hart Hotel, and the assistant battalion commander, with the rank of captain, was colliery official W. H. Sansom.[28] The previous month, February 1941, it had been the turn of the Special Constables of Ollerton to dine, under the presidency of the chief special constable—none other than Montagu Wright of Ollerton Hall.[29] In 1942 the Boys Brigade in Edwinstowe formed an army cadet force, and the Girls Brigade a girls' training corps, although the latter's numbers were diminished by the blackout and by the girls' employment in the Mansfield munitions factories.[30] A bomb disposal squad was formed in Edwinstowe in 1941.

Besides the physical preparations the Dukeries coalfield undertook as a result of the war, propaganda designed to stiffen morale was to be found in abundance. When a new canteen for the miners at Firbeck Colliery was opened in January 1943, the *Worksop Guardian* reported that as a result, 'the miners can better fight Hun, Wop and Jap'.[31] In April 1942 Butterley Company collieries general agent

[24] Butterley H3 224.
[25] Thoresby Colliery Manager's Annual Report, 1941.
[26] Ibid.
[27] Ibid., 1941.
[28] *Ollerton, Edwinstowe and Bilsthorpe Times*, 8 Mar. 1941.
[29] Ibid., 28 Feb. 1941.
[30] Thoresby Colliery Manager's Annual Report, 1942.
[31] *Worksop Guardian*, 22 Jan. 1943.

Montagu Wright ordered John Bull posters and posters reading 'It All Depends On Me', and announced to colliery managers that their purpose was 'to spread amonst your employees the spirit of Individual Responsibility', on sending a selection for display at each of the Butterley Company's pits.[32] Later that year 'It All Depends on Me' stamps produced by the Universal Postal Frankers of London were used by Butterley, because 'well-known and patriotically minded users' would help to bring an early victory, since 'in a democracy it is the combined effect of individual effort, more than any more Government machinery, that will bring us an early victory'.[33] In 1943 a Ministry of Information photographic and photogravure display for propaganda purposes was exhibited at Ollerton Colliery.[34] Needless to say, the coal companies felt that a patriotic stimulus which improved output would help their own industrial relations and profits as well, but in many ways the miners and their leaders had other ideas as to how the war might be used to improve their industrial position.

Like all other coalfields, the Dukeries experienced in the latter part of the war an influx of Bevin Boys from outside the industry. At Thoresby, fifty-five Bevin Boys began work during 1944, and twenty-one more in 1945.[35] The miners' attitude towards the Bevin Boys exhibited a not unexpected dualism:

HT They took them under their wing. They were a bit of a joke to them, but if you know what I mean in a nice sort of a way. They made friends and they went to the pub with them.

Mrs H I can remember our Frank going off pop about the Bevin Boys that he got and at the same time nursing them along and looking after them underground, you know.[36]

Generally the Bevin Boys were accepted by the miners as long as they weren't 'stand-offish', but very few stayed on in the Dukeries coalfield after the war; one notable exception was Frank Haynes, originally from London, who came to

[32] Butterley P9 292 'Propaganda'.

[33] Ibid.

[34] Ibid.

[35] Thoresby Colliery Manager's Annual Reports, 1944–5.

[36] Interview, Hilda Tagg, Mrs Howard, Rainworth.

Thoresby Colliery and was eventually elected MP for the Ashfield division of Nottinghamshire.[37]

Most of the effects of the Second World War described above have much in common with the experience of many parts of Britain. But two consequences remain which place a very different light on the story. These developments concern the accession of prosperity and political power which the full time and essential production of wartime brought to the coal miners of the Dukeries. First, there can be little doubt that wages rose substantially in the new Nottinghamshire coalfield during the war:

RJW I was asking if you can remember how the war affected Rainworth, and what kind of changes took place then?

HT Well, it brought the first bit of money to Rainworth, I think, because once things got into full swing, there were full-time employment and overtime, as opposed to short time and poor work and so on. You see from doing three days a week I suppose by the first year of the war they could have taken the best of the pit if they'd wanted to.[38]

This subjective evidence is overwhelmingly corroborated by the statistics for days worked, production output, and wages. Output at Thoresby Colliery increased from 647,127 tons in 1938 to 908,456 in 1944. With the help of the Greene Committee and the Porter Award of 1944, wages per getter per shift at Thoresby increased from 20*s.* 5*d.* in 1939 to 41*s.* 9*d.* in 1945.[39] The number of days worked at Ollerton Colliery rose from 186¼ in 1938 to 295¾ in 1943.[40] All in all, it is clear that the war offered higher money wages to mining families, even if shortages took some of the gloss off the improvement.

Second, there can be no doubt that the miners' political position improved as a result of the Second World War. After many years of quiescence, ominous signs of recalcitrance began to appear in the section of the Thoresby Colliery Manager's Annual Reports entitled 'Discipline'. In 1940 the manager reported that 'much greater efforts have to be made

[37] Correspondence, Frank Haynes. Haynes was elected MP for Ashfield in May 1979.

[38] Interview, Hilda Tagg, Rainworth.

[39] Thoresby Colliery, Manager's Annual Reports, 1939–45.

[40] Butterley C4/2 33, 'Ollerton Colliery 1942–6'.

to tactfully enforce discipline especially the timbering rules underground'. In 1941 it was reported that 'Discipline has been maintained fairly well but the Deputies have not the same hold on the men as they used to. The timbering rules have had to be continually but tactfully enforced. People do not seem to have the same pit sense as they used to.' By 1943 it was becoming clear that the company was fighting a losing battle. The report on Discipline begins:

I am afraid that this has deteriorated although we have had no serious trouble. The men have got the upper hand and they know it. I am sorry to say that some officials have not done what they might do to maintain discipline. Trouble seems to be simmering in the background all the time and it only wants a small thing to set it off. We are promised some sort of restoration of discipline by Major Lloyd George but even if we are given more power this may not by itself secure that goodwill which ought to exist between the employers and their employees which is so essential, and indeed a necessity, to good production, order, and efficiency.

The Government appear to be strict enough with the employers, but do not enforce discipline on the miners.

We dismissed, for gross misconduct, 14 workmen during 1943.[41]

Nor were problems of indiscipline confined to Thoresby. At Rufford, once a stronghold of Spencerism,[42] the colliery manager W. V. Sheppard reported in 1944 that

Although no stoppage of work has been caused in this, or any other War Year, there has again been an increased influence from a growing section of the workmen addicted to Communistic tendencies. The original members of this section of the workmen have steadily spread their creed until there exists a strong minority obviously opposed to the Company and all it stands for. Without such organised agitation I have no doubt that results would improve at once.[43]

The main source of complaint was the steadily increasing absenteeism registered during the war years. At Clipstone this rose from 8.4 per cent in 1940 (only 3.1 per cent of which was classed as 'avoidable absenteeism') to 19.9 per cent in 1945.[44] At Ollerton, absenteeism rose from a level of 5.96

[41] Thoresby Colliery Manager's Annual Reports, 1943.

[42] See above, ch. 5.

[43] Bolsover Company, Rufford Colliery Manager's Report to Managing Director, 1944.

[44] Bolsover Company, Clipstone Colliery Manager's Annual Reports to Managing Director, 1940-5.

per cent in 1935 to 20.94 per cent in 1945.[45] At the coal face the figure reached an all-time high for the colliery at 32.98 per cent in July 1945.[46] Company internal correspondence makes no bones about the chief reason for high absenteeism: deprived of the threat of dismissal by the Government employment restrictions, the companies could no longer dominate their employees as they had since the opening of the pits and throughout the pre-war period.[47] One interviewee from Harworth pointed to the war as the beginning of the end for company control in the villages:

NH It fizzled out from the day almost when the war came along. The war put certain restrictions on colliery managers. You see previously if he didn't like anybody, if he took an intense dislike to someone, because they didn't always see his way, he'd just sack them. He'd just sack them. But because the war came, under the emergency powers, they couldn't sack anybody at the pit.

RJW So you felt that during the war things were changing, because although the pit wasn't really affected, say wasn't bombed, say, and men didn't go away to fight, because they were miners, nevertheless it was a time of change?

NH Yes, yes. A time of change, I think it altered the whole history and the attitudes of people. It seemed to me that men who were pacified to continue apathetic in the old way that they were bullied before, the end of the war changed their attitudes, you know, even the simplest of persons, 'oh, no, they're not going to bloody talk to me like that no more, I'm going to have a say from now on', and they did! It did alter people's attitudes.[48]

It was hard enough even to change pit voluntarily in the war years, regardless of the new-found rarity of dismissal. When Wilf Sperry wished to move from Bilsthorpe to Ollerton in 1942, he had to convince government officials that he had good reasons for wanting to make the change.[49]

It is no accident that the war years also saw the relaxation of several other restrictions imposed in the company villages of the Dukeries coalfield. In December 1941, the miners of Ollerton were allowed to keep greyhounds for the first time.[50]

[45] PRO POWE 35/246.
[46] Butterley M2/1 252.
[47] For alternative explanations see A. R. Griffin, *Mining in the East Midlands 1550-1947* (London, 1971), pp. 287-8.
[48] Interview, Neville Hawkins, Harworth.
[49] Interview, W. Sperry, Ollerton.
[50] Butterley H3 224, 18 Dec. 1941.

For the first time a Communist was allowed to speak in the Butterley village in 1943, in favour of a united front candidate at the Newark by-election.[51] The foundation of the Labour party in the new Nottinghamshire coalfield in the war years has been described above.[52] In addition to the specific effects of the diminution of company power in the colliery villages, the deep and radicalizing effect of the war apparent throughout Britain should be borne in mind. The Labour party contested parish and district council elections for the first time in the new mining villages in 1946 and swept the board.[53] It is not surprising that one of the first Labour party members in Bilsthorpe remembered the political developments as the most notable consequences of the war:

RJW I'd like to ask you what effect the war had on Bilsthorpe, what changes there were in the village?

HT I would say very little, excepting politically of course, the political upheaval from one side to the other which has been maintained ever since.[54]

The war, more than anything else, broke the power of the coal companies in the Dukeries mining villages. Freed from the threat of arbitrary dismissal and placed in the position of producing a commodity essential to the war effort, the miners were able to organize industrially and politically for the first time, and a social revolution in the villages was made possible by the current of opinion flowing towards nationalization of the industry. It is interesting that given the opportunity, the majority of Dukeries miners and their wives seemed to wish to reject the trappings of coal-company control. This evidence suggests that most had resented the situation that had pertained before the war, and that their political inaction can be explained more through their powerlessness than through positive assent to the ideals of the 'model' village, harmonious and socially integrated.

Nationalization: 'That's when the rot set in'[55]

Just as in the case of war, the effects of nationalization on

[51] Interview, A. E. Corke, Ollerton.
[52] See above, p. 152.
[53] See above, pp. 160–2.
[54] Interview, Herbert Tuck, Bilsthorpe.
[55] Interview, George Cocker, Edwinstowe.

the Dukeries coalfield were not those which might at first be expected. Few interviewees look back on Vesting Day as a moment of great release or liberation from company control, and it is doubtful whether there really was a dramatic shift in power in favour of the miners at the expense of the employers. Nevertheless, the companies did themselves have much to lose, and the Dukeries coal-owners took a lead in fighting the proposal to nationalize the most profitable coalfield in the country.

The wartime file in the Butterley Company's archives marked 'Propaganda' contains material relating both to patriotic morale, and, after 1943, to the company's campaign to resist post-war nationalization. In January 1943 we find a memorandum to the following effect:

> It is the considered view of the Midland Counties (Coal Owners Association) that a Central Publicity Organisation shall be instituted immediately with the following objects:
> 1. To be prepared to deal with any contingency as and when it arises and in particular—Nationalisation post War.
> 2. To counteract Communistic propaganda in the coalfields.
> 3. To present the Coal Owners point of view to the general public in a more favourable light, particularly in non-mining areas.
> 4. The question of parliamentary representation is of the utmost importance.
> 5. Advisability of obtaining the services of a first-class journalist.
> 6. It is not suggested that until the necessity arises the organisation should indulge in any propaganda hostile to the MFGB.[56]

All these suggestions were adopted by the Dukeries coal companies during the last four years of their ownership of the mines. To combat Communism the Butterley Company employed Stan Middup, formerly a Spencer-union official, to act as a kind of secret agent helping 'moderate' candidates to be elected as checkweighmen and union secretaries. When Dai Ley looked like being chosen as checkweighman at Kirkby, Middup was asked by the Propaganda Committee 'to get the men interested in some of the other candidates'.[57] Attempts were made to prevent the Yorkshireman Sam Kilner, founder of the Labour party in Ollerton, from being elected

[56] Butterley P9 292 'Propaganda'.
[57] Butterley P9 292 'Propaganda', 27 Feb. 1943.

checkweighman at Ollerton Colliery in 1943.[58] Communist tracts such as *The Coal Crisis* (1943), which called for state control of the mines and pointed to the coal-owners' political activities and profits, were answered by material such as the fifteen Mining Association broadsheets of 1945 and J. P. Dickie's *613 Questions about the Coal Industry.*[59] As suggested in the midland coal-owners memorandum, a journalist wrote a series of articles designed to publicize the coal-owners' case against nationalization: J. C. Johnstone of the *Daily Telegraph* produced articles on the Nottinghamshire coalfield on 29 December 1944 and 13 March 1945. The well-known Alfreton journalist W. Smithurst performed the same task at local newspaper level—he had been on the staff of the *Derbyshire Times* since 1896, but died suddenly in December 1944, at the age of 70.[60] But despite the efforts of the Butterley Company and its allies at every level from Parliament to pit, the cause was hopeless, and the last item in the publicity file is a sad little note of July 1946, asking if the Publicity Committee might as well be wound up. The coal companies logically fought the nationalization of the Dukeries mines: at the end of the war their profitability was still keeping some businesses afloat. When the Butterley Company's coal venture went into liquidation on 28 June 1946, it could report a profit of £188,000 for the last year of its operation—still, apparently, a flourishing concern.[61]

How did the miners and their families see the impact of the public ownership of the mines? Certainly it helped to consolidate the gains in job security and political power won during the war: 'It meant that gone were the days when a man turned up for work and he was told, "Oh, no, there's no work for you today," so of course it meant that he'd suddenly got a steady job, and sort of a regular job.'[62] A Bolsover Company deputy thought that it was harder to enforce discipline without the threat of instant dismissal:

RJW What effect do you think nationalization had on Edwinstowe?

[58] Ibid., 1 July 1943 'Re Kilner Ollerton'.

[59] Ibid., June 1945.

[60] Ibid., Dec. 1944.

[61] R. H. Mottram and C. Coote, *Through Five Generations: the History of the Butterley Company* (London, 1950), p. 33.

[62] Interview, Marjorie Wilson, Mansfield (Blidworth).

JHG Well, I don't think it had a great effect on Edwinstowe as a whole, but on the pit, yes, it did, because during that period you see I'd got onto the staff in '37, so I'd been a deputy under the Bolsover Company for ten years, and as such we was backed right up to the hilt with the management. For instance, if I had got a man that weren't really pulling his weight, and I said 'take your tools, you're finished here', that was it. He'd finished. Now of course under nationalization you can't do that.[63]

Things in the village *did* change, as well. No longer was the colliery manager unquestioned head of the colliery community hierarchy. Indeed often the NCB managers did not live in the villages. Yet despite the disappearance of the coal-owners like Montagu Wright of Ollerton Hall, the impression remains that for many inhabitants of the Dukeries colliery villages, one set of bosses was merely replaced by another set: 'Not a great deal of difference. I mean we knew that the work had got to be done, and carried on just the same only lived by another name, sort of thing.'[64] Many respondents indicated that in their opinion nationalization had brought no revolution to the life of the miner. What is more, for some the change was actually a backward step. There was still no feeling that the miner was in control of his own life, or doing anything other than a hard and dangerous job for little reward. Stripped of the paternal interest of the coal companies, the villages began to look as if no one took any pride in them:

RJW You said earlier that the company did take an interest in what went on in Edwinstowe village, kept it neat and like a model village. Do you think that changed after the war?

JHG Well, it got as nobody bothered. It got as nobody bothered, you see, after. Some people allowed their gardens to get overrun.[65]

Comparisons between the days of company control, for all their faults, and the period since nationalization are by no means as clear cut as might be thought:

RJW Presumably all the houses in the colliery village were once owned by the Bolsover Company but now they were owned by the NCB. Did this mean that some of the regulations like keeping the gardens tidy were relaxed?

[63] Interview, J. H. Guest, Edwinstowe.
[64] Interview, J. E. Smith, Ollerton.
[65] Interview, J. H. Guest, Edwinstowe.

GC That's when the rot set in.
FT Vesting Day was the day when the rot set in for the village.
RJW So you think that the village looks now a lot less tidy and neat?
FT Oh, it's scruffy. Scruffy compared to those days.
GC This was the neatest colliery village in the country then.
FT Yes, it was a model. It was a model village in the country.[66]

Even amongst pioneers of the Labour party in the Dukeries coalfield, a favourable view of the model villages under private ownership can be found:

Ollerton itself as a village was a beautiful place, no hooliganism, no nothing of that sort, everything contained in itself, sort of thing. They'd a decent football team, a decent cricket team, a band perhaps, a male voice choir and all this sort of thing . . . the colliery village at Ollerton was the best colliery village I've seen in my life, and I've seen a few, for its layout and everything.[67]

One could possibly describe attitudes such as these as representing nothing more than the success of the private coal-owners in imposing their values and norms on the inhabitants of the colliery village, but the fact remains that the orderliness and neatness of the model villages genuinely appealed even to those who were opposed to private control of the industry. Many of those who remember the colliery villages before nationalization have mixed feelings about its effects:

RJW What did the end of company ownership mean to the village?
CES Well, it meant freedom. They got rid of the taskmasters. They were free then, and they decided what they should do.
RJW What happened to all the little regulations about keeping gardens tidy and so on?
CES Oh, they all went by. It had its bad side. I've never seen the village as dirty as it is now. Yes, the village has never been as neat, as tidy, as clean, as what they did, by enforcement. There's no doubt about it, no one disputes that. They kept it lovely. They argued that, and they did. And it was that enforcement that made them do it. Whether they liked it or not, they had to do it. And all these plantations, all these streets, were beautiful. Now you find just stones. There wasn't a cigarette end or anything. They have freedom now, but they pay for it, because it's not as beautiful.[68]

The end of the coal companies meant the end of some of the most blatant abuses of power, such as arbitrary evictions

[66] Interview, George Cocker, Frank Tyndall, Edwinstowe.
[67] Interview, W. Sperry, Ollerton.
[68] Interview, Charles Stringer, Harworth.

and sackings, and the effective prohibition of militant political and industrial activity; and in this sense nationalization could be seen as a step forward for the Dukeries. But it would conflict with the evidence to pretend that nationalization meant 'public control' in anything but name. The miners in general still felt that they were tied to their jobs by little more than the cash nexus; that management was still 'them'; and that their interests still could not be identified with those of the industry as a whole. Nationalization removed the paternalistic employer with their rigorous scrutiny of the miner's life in the colliery and in the village. But few were able to see it as a revolution which opened up many bright new opportunities for the miner and his family.

If the character of the Dukeries colliery villages has been changed since the war, it has been far more due to the continued growth of the communities, and the arrival of alternative forms of employment, than to the nationalization of the coal-mining industry. At Harworth the colliery is no longer even the largest single employer in the village. The Glass Bulbs factory employs 1,650 workers compared with the pit's 1,150, and there is also an Airfix shoe factory which caters mainly for female employment. In the Ollerton-Boughton community, despite the fact that most of the miners employed at Bevercotes Colliery (sunk between 1953 and 1958) live in the village, a new industrial estate has diluted its character as a colliery village. At Langold, the pit closed in 1968, and an industrial estate of half a dozen factories erected on its site. Many men were re-employed in the mining industry and travel from Langold model village to Manton, Maltby, or Shireoaks collieries, but a pit-village without a pit remains an oddity, especially as far as younger people starting work are concerned.

The nature of the colliery villages of the new Nottinghamshire field has also been much altered by the recent NCB policy of allowing tenants other than their employees to live in their houses. Indeed in most of the villages, plans are in operation to sell as many of the NCB houses as possible, something the coal companies would never have considered. Local authority housing estates have been built since the war in all the Dukeries colliery villages. With the growth of car ownership, many people now travel from the mining villages to work in Mansfield or further afield. In fact the old villages of Ollerton, Edwinstowe, and Bilsthorpe have themselves grown quickly in the 1960s and 1970s as they have become fashionable commuting bases for the middle classes, for whom extensive private housing developments have been constructed, such as that known by the miners as 'Beverley Hills' at Edwinstowe. Similarly, it is now possible for men working at the Dukeries collieries to drive in from homes ten or twenty miles distant. The old isolation of the colliery village, the old connection between pit and pit village has been irretrievably broken.

It is difficult to ascribe this change to the policy of nationalization, although the private coal companies might have attempted to resist the assault on the principle of the planned one-industry colliery village, built and controlled by one authority in common with the mine. But that principle could not have been maintained in a post-war world of mass transport and mass communication. One resident of Ollerton pointed out the transformation with possibly unconscious symbolism: 'Nobody was supposed to walk on the lawns, where you can see cars on them now. And there used to be a chap, we used to call him Bobby Healey, and if he caught anybody walking on the lawns, he used to prosecute them, he used to fine them at the pit.'[69] As at Saltaire and Bournville, the regular plan and layout of the Dukeries colliery villages still stand as memorials to the initiative and ambition of the private coal companies who built them. The cars which everywhere stand in the streets and on the lawns on front of the garageless houses emphasize the outmoded nature of that initiative, and perhaps cast doubt on the long-term success of any rigorously planned community in the face of economic and technological change.

[69] Interview, G. Smith, Ollerton.

11. The Problems of Developing Coalfields

'Entangled within a morass of social ills . . .'[1]

The early history of the Dukeries coalfield was typified by dislocation, rather than by the social stability often associated with mining areas. A very high rate of labour circulation and an unusually complete local domination by the coal companies meant that miners did not become as homogeneous a body socially or politically as might be expected. For many years their lives were disrupted by inadequate services, poor educational arrangements, a lack of employment for female labour, and villages resembling private building sites rather than established mining communities. All this contrasts sharply with Martin Bulmer's sociological model of a mining community as marked by a pride in group activity outside the pit as well as underground, a low level of geographical mobility, and strong industrial and political organization.[2] Yet if the new Nottinghamshire coalfield were merely an exceptional case, its study would be of little more than antiquarian value. Perhaps its lessons and parallels give it wider significance.

First, it would seem that the character of the Dukeries coalfield in the inter-war years was in many ways typical of the early stages of development of other British coalfields. As Arthur Smailes put it in his article in the *Geographical Journal* for 1938:

> Mining is a form of destructive exploitation of the earth's resources, a fact which is reflected by the population cycle of colliery settlements. The stage of youth, immediately following the commencement of mining, is a period of immigration and rapid increase of population.[3]

[1] Social Services Department, Rushcliffe Council. *Cotgrave* (1977). See below, p. 244.

[2] M. I. A. Bulmer, 'Sociological Models of the Mining Community', *Sociological Review*, NS 23 (1975).

[3] A. E. Smailes, 'Population Changes in the Colliery Districts of Northumberland and Durham', *Geographical Journal*, 91 (1938), 220-32.

Smailes goes on to describe how as the coalfield passes into maturity, immigration falls off as sufficient new recruits are supplied by the community itself, and then the stage of maturity itself passes as further development of the pit or field cannot provide sufficient work, and outward migration commences, and a new field must restart the cycle.

Smailes's example of the demographic cycle of a mineral extractive industry is the history of the North Eastern coalfield in the nineteenth century. The oldest coal producing area, around Tyne Bridge, had been exploited for centuries but was declining by 1800, replaced by new mining further east as new machinery enabled deeper coal to be reached, for example after the sinking of the pit at Wallsend in 1780. Between 1801 and 1831 there was a 60 per cent increase of population in Durham County, but the real rise due to immigration did not come until after the opening up of inland Durham by the construction of the Stockton and Darlington railways after 1825. Much immigration from other counties took place between 1825 and 1850. New pits were sunk near to the new coastal ports. As the report of the Child Employment Commission of 1841 put it:

> Within the last ten or twelve years an entirely new population has been produced. Where formerly there was not a single hut of a shepherd, the lofty steam engine chimneys of a colliery now send their volumes of smoke into the sky, and in the vicinity a town is called, as if by enchantment, into immediate existence.[4]

After the 1880s another period of growth was ushered in as the Low Main seam was exploited and new shafts sunk in south-east Durham and the Ashington area behind the Northumberland port of Blyth. However by the time Smailes wrote in the 1930s the cycle had reached another stage. The decline of the North Eastern coalfield had resulted in mass unemployment and outward migration, both from the industry itself and to newer coalfields like that of the Dukeries. In the words of R. S. Moore:

> The changes in the economics and structure of the coal industry have been reflected in County Durham. The earliest pits were close to the

[4] Report of the Child Employment Commission, 1841, Appendix 1, p. 143.

surface, seeking easily won coal in the drift mines in the west of the county. The late nineteenth century boom created the pits of mid-Durham, deeper pits with shafts and more extensive haulage equipment . . . in the 1960s the coal industry in the county was being concentrated in the very deep and highly-mechanised pits of the east coast.[5]

Moore related this development to the rise of the large corporations and highly rationalized production of the twentieth century, changes accompanied in his view by considerable industrial and political conflict. A steady eastwards migration took place, from Weardale to the Deerness Valley, from Deerness to the coast.[6]

It is clear that demographic change was a concomitant of the development of many British coalfields. The immigration consequent upon the opening up of a new area was followed by a period of stability and then a decline as pits were forced to close by the exhaustion of mineral resources. Some cases from other coalfields pre-dating the Dukeries may be briefly mentioned. For earlier periods the evidence necessary for a detailed social history of young coalfields is unfortunately scanty, but large-scale migration took place as far back as the early nineteenth century when the history of mining in Ayrshire[7] reveals that the industry moved as the exposed seams were exhausted, and the development of transport in the region opened up the possibility of exploiting new fields further inland. A tramway was built around 1810 from Ayr to the mines at Whitletts, and the Troon-Kilmarnock tramway (1812) and the Ardrossan-Kilwinning railway (1827) followed. The ports at Saltcoats, Irvine, Ayr, Ardrossan, Troon, and Girvan expanded during the nineteenth century with the opening of new mines and the subsequent stimulation of trade. Then in the early twentieth century the greatest shift occurred, as old, small pits closed and were replaced by new, deep pits in the Dalmellington area, New Cumnock, and the concealed Mauchline basin. Production levels remained virtually constant overall throughout the period from 1894 to 1928, but the location of industry changed, as shown in Table 36.

[5] R. S. Moore, *Pitmen, Preachers and Politics* (Cambridge, 1974), p. 31.

[6] Ibid., p. 66.

[7] See J. H. G. Lebon, 'The development of the Ayrshire coalfield', *Scottish Geographical Magazine*, 49 (1933), 138-53.

Table 36. *Employment in the Ayrshire coal industry, 1894–1928*[a]

	% of men employed	
district	1894	1928
Dalry	6.0	1.0
Kilmarnock basin	41.0	27.0
Mauchline basin—exposed	28.0	10.0
Mauchline basin—concealed	1.0	29.0
Muirkirk basin	8.0	8.0
Dailly basin	1.0	2.0
Dalmellington	7.0	11.0
New Cumnock	7.0	12.0

[a] Lebon, 145.

In undertaking a study of two contrasting parts of the booming Lanarkshire coalfield in the mid-nineteenth century, Alan Campbell considered the reasons for the relative lack of trade union activity in Coatbridge compared with the area round Larkhall.[8] He discovered that Coatbridge had a heterogeneous character, a large number of Irish immigrants being employed. The Irish fought with the Scots, acted as a labour surplus, intermarried but rarely with local girls, and (despite their riotous reputation) rarely organized stable and enduring unions or provided many labour activists.[9] Sectarianism weakened labour solidarity, and the Irish population proved highly geographically mobile. Secondly, the employers in the Coatbridge area tended to be larger than those in Larkhall, often diversifying iron manufacturers. Often they formed a local élite of family dynasties.[10] The large companies were in a better position to dominate property ownership, providing a large amount of company housing which was strictly controlled, and monopolizing political functions like the magistracy and school boards. Company stores were common in Coatbridge, and the truck system could not be effectively eliminated.[11] Like the Dukeries companies, the Lanarkshire employers were often bitter rivals, but this did not stop them co-operating against organized activity amongst the miners.

[8] Alan Campbell, *The Lanarkshire Miners* (Edinburgh, 1979).
[9] Ibid., p. 195. [10] Ibid., p. 209. [11] Ibid., p. 218.

The 'social control' exercised by the companies was not necessarily a conscious, hypocritical, and instrumental policy —but its effects were undoubtedly significant in reducing trade union activity in the Lanarkshire coalfield.[12] In general, Campbell concluded that 'the greater the degree of occupational differentiation, of geographical mobility, and of industrial concentration, the weaker trade unionism among the colliers should be'.[13] 'The rootless and heterogeneous population in the mushrooming mining settlements on Lanarkshire's industrial frontier, the ethnic and religious fragmentation brought about by the coming of the Irish, and the modes of social control developed by the iron companies acted as a series of barriers to the establishment of stable unions in the coalfield during the middle decade of the nineteenth century.'[14] Although Campbell limited himself to the causes of dislocation of trade union activity among the miners, the parallels with the new Dukeries coalfield of later years are clear.

In the coalfield around Cannock Chase too, there exists a contrast between those areas opened up in the nineteenth century (Cannock parish's population increased from 3,081 in 1851 to 26,012 in 1901) and the newer concealed field. Featherstone parish increased from thirty-nine in 1921 to 1,058 in 1931 due to the opening of Hilton Main Colliery in 1924.[15] One may note some social consequences which may be gleaned from the demographic evidence. In Cannock Urban District in the twentieth century there was a serious female deficiency in the fifteen to twenty-four age group as a result of girls being forced to migrate to seek jobs off the coalfield. In the 1931 census there were 2,213 females between fifteen and twenty-four years of age, and 3,338 men. In Brownhills in 1947 80 per cent of the insured population were miners, and only 9 per cent of the insured population were female—a figure which may be compared with 27 per cent in the nearby cathedral town of Lichfield.

Population studies offer us a limited but quantitative guide to the pace and timing of economic change in localities. For

[12] Ibid., p. 205. [13] Ibid., pp. 234–5. [14] Ibid., pp. 2–3.

[15] M. J. Wise, 'Some notes on the growth of population in the Cannock Chase coalfield', *Geography*, 36 (1951), 235–48.

example, C. E. Redmill contrasts, by means of parish studies, the rise of mining in East Warwickshire with the decline of agriculture in this previously rural area.[16] A rapid growth of population may be noted in the parishes of Arley (1901–11) and Binley (1911–21) while other parishes went through a complete population cycle, declining once more as the local pit closed. Another interesting area was Shropshire, for in the Darby works may be seen a pioneering stage in the Industrial Revolution itself, yet Shropshire was left behind by its rivals in the nineteenth century, partially at least due to transport difficulties. Population figures for the parishes of Priorslee, Broseley, and Little Wenlock illustrate the demographic effects of the rise and fall of the coal, iron, and clay industries in the nineteenth century, a changing pattern as mineral resources were exhausted.[17]

It is clear that in Nottinghamshire itself the opening up of the Dukeries coalfield in the 1920s was but part of a series of developments as the older seams of the exposed coalfield of Derbyshire and Western Nottinghamshire gave way to the new pits of the concealed coalfield: the Leen Valley had been exploited for the first time in the 1860s and 1870s, with the sinking of such pits as Hucknall nos. 1 and 2 (1861–2), Annesley (1865), Bulwell (1867), Bestwood (1871-2), Linby and Watnall 1 (both in 1873), and Newstead (1875). Around the turn of the century, the district around Mansfield was developed: Sherwood (1902-3), Mansfield (Crown Farm) (1904-5), Rufford (1912-13), and Welbeck (1915).[18] Lord Taylor (formerly Bernard Taylor, MP for Mansfield from 1941 to 1966) was born in Mansfield Woodhouse, then a rural village on the edge of Sherwood Forest, in 1895, and in his autobiography he reports how this district, previously dependent upon agriculture, stone quarrying, and textiles, was transformed by the sinking of Sherwood and Crown Farm collieries.[19]

[16] Cecil E. Redmill, 'The growth of population in the E. Warwickshire coalfield', *Geography*, 16 (1931), 125-39.

[17] T. W. Birch, 'Development and decline of the Coalbrookedale coalfield', *Geography*, 19 (1934), 114-26.

[18] A. R. Griffin, *Mining in the East Midlands 1550-1947* (London, 1971), pp. 160-81.

[19] Lord Taylor, *Uphill All the Way* (London, 1972), p. 1.

Migration and demographic change is inherent in the development of the coal industry. It is hard to find evidence for the social effects of the opening of new pits in previous centuries, but there is at least subjective evidence for a high degree of social dislocation and confusion. Shirebrook in Derbyshire was rumoured to be a real 'frontier town' around the turn of the century;[20] and we possess Jack Lawson's reminiscences of life in the new mining town of Boldon in eastern Durham in the 1890s:

> Boldon Colliery was at that time a typical example of the way in which the county of Durham had become a sort of social melting pot owing to the rapid development of the coalfield . . . its population consisted of people from every part of the British Isles, some of the first generation and some of the second . . . at the time of which I write there was a combination of Lancashire, Cumberland, Yorkshire, Staffordshire, Cornish, Irish, Scottish, Welsh, Northumberland and Durham accents, dialects and languages. While we were all good neighbours, I have seen the clans come together and fight it out in very rough and ready style. If anyone thinks the blood bond does not matter, let them live under such conditions, and their theory will be strained, to say the least of it.[21]

It would certainly seem that a high turnover of population together with social and economic disruption were a common feature of the early stages of the exploitation of British coalfields in the nineteenth and twentieth centuries alike. But it is not clear that nineteenth-century colliery villages were influenced by the coal-owners to the extent that the new Dukeries mining villages of the 1920s and 1930s were. Few studies have analysed the early days of the nineteenth-century colliery communities. Robert Moore showed that several of the colliery villages constructed in the Deerness Valley west of the city of Durham around 1870 were 'company towns', every brick in the colliers' houses being stamped with the name of the coal-owner.[22] The owners and management were strongly paternalist. Here too, as in the Dukeries, the companies donated money for chapels and churches, constructed schools, encouraged flower shows, libraries, and sport, inspected the miners' gardens, and sponsored dances

[20] J. E. Williams, *The Derbyshire Miners* (London, 1962), p. 464.
[21] Jack Lawson, *A Man's Life* (London, 1932, 1944), p. 36.
[22] Moore, *Pitmen*, p. 81.

and other social events. Company officials sat on the local councils.[23] Moore believed that the paternalist approach was killed by the economic pressures of the decline of the coal industry in the twentieth century, which made it hard to maintain an image of harmony in the villages when profits and wages were both being squeezed. This culminated in the disastrous 1926 strike, which ended for ever the co-operative spirit which had undoubtedly existed in the nineteenth century.[24] In mid-Durham we can see how the opening up of a new field could lead to an initial period of company control.

However, Philip N. Jones concluded that in the case of South Wales, at least, the development of the coalfield was piecemeal, unplanned, and not conducted under the auspices of a single authority in each locality.[25] Few mining villages in South Wales could be described as carefully planned communities on the lines of the 'model villages' of Nottinghamshire. For a start, the companies did not provide all the housing in the South Wales villages. The Royal Commission on Coal of 1873 found that 'housing in the coalfield involved the coal-owners to only a very minor degree'.[26] The provision of houses by the owners depended to a large extent on the relative isolation of the pits from established communities. Few dwellings were supplied by coal-owners in Lancashire, where the collieries were sunk in populous areas such as that near Oldham,[27] and in Staffordshire,[28] but 'comfortable accommodation' was provided in Yorkshire,[29] Ayrshire,[30] and Durham, where J. W. Pease testified to the necessity of building houses as an inducement for miners to come into the district.[31]

Many joint-stock building companies and speculative jerry-builders helped to supply housing for the labour migrating into the South Wales coalfield in the second half of the

[23] Ibid., pp. 79-86. [24] Ibid., p. 92.
[25] P. N. Jones, 'Colliery Settlement in the S.Wales Coalfield, 1850-1926', University of Hull, *Occasional Papers in Geography*, 14 (1969).
[26] PP 1873 x, Report of the Select Committee on Coal, p. 91.
[27] Ibid., p. 115, evidence of I. Booth.
[28] Ibid., p. 256, evidence of W. Brown.
[29] Ibid., p. 140, evidence of R. Tannant.
[30] Ibid., p. 300, evidence of A. Gilmour.
[31] Ibid., p. 190, evidence of J. W. Pease MP.

nineteenth century. Often the houses provided were of poor quality. Some coal companies did build houses for their workmen, especially between 1899 and 1921, but there is little suggestion of the monolithic control apparent in the later colliery villages which could be 'planned from scratch'. After 1919 the Labour-controlled local authorities in South Wales could build council houses in far greater numbers than, say, Southwell Rural District Council.

Another reason why the coal companies were less able to intervene in many aspects of miners' lives in the nineteenth century may have been that the ownership of mines tended to be divided among small proprietors. Raphael Samuel argued that a rise in supervision of employees followed the creation of large companies like the Cambrian Combine and the Powell Duffryn group in South Wales: 'The later years of the nineteenth century saw much more aggressive forms of capitalist intervention, and mineral workers were affected by it no less than, say, engineers . . . the small proprietor was eliminated.'[32] It does seem likely that it would have been harder for the coal companies of the new Dukeries field of the twentieth century, had each mining community been based on several small pits each employing a handful of workmen and each owned by a different proprietor.

Parallels may also be drawn between the history of the new Dukeries coalfield and other developments in the British coal industry in the twentieth century. The South Yorkshire coalfield around Doncaster overtook the longer-established West Yorkshire field as far as production was concerned in 1913 due to the sinking of deeper mines in the first few years of the century. Here too, as in Nottinghamshire, there was a general tendency for miners to migrate from west to east as new, deep pits were sunk in the concealed coalfield.[33] The new, eastern pits were larger, as Table 37 indicates. In the NCB division containing the oldest pits in Western Yorkshire, Barnsley, most of the pits employed fewer than 500 men. In the more recently developed Rotherham area

[32] Raphael Samuel (ed.), *Miners, Quarrymen and Saltworkers* (London, 1977), p. 74.

[33] Chrisopher Storm-Clark, 'The Miners 1870–1970, a Test Case for Oral History', *Victorian Studies*, 15 (1971–2), 51.

Table 37. *The Yorkshire coalfield, 1971*[a]

	eastwards - -> concealed - -> developed later - ->		
collieries	NCB area: Barnsley	Rotherham	Doncaster
0–499 men	13	1	0
500–999 men	2	10	0
1000–499 men	3	5	3
1500–999 men	3	3	5
2000+ men	1	0	2

[a] K. Peace, 'Some Changes in the S. Yorkshire Coalmining Industry', *Geography* 58 (1973), 341.

most of the pits were somewhat larger, mostly employing between 500 and 1,000 men. In the most recently developed, eastern field around Doncaster the undertakings were significantly larger, half of them with between 1,500 and 2,000 miners.[34]

It was recognized at the time that the stakes for which the coal companies were playing in the Doncaster field were high. As H. S. Jevons wrote in 1915:

> The future extension of the coalfield must be to the east, to the unproved coalfields to which reference has already been made. Great developments are already taking place to the east and south of Doncaster, and it is anticipated that this portion will become the wealthiest of the new coalfield.[35]

South Yorkshire, together with the Dukeries, was to become the mainstay of the twentieth-century British mining industry. Brodsworth Colliery reached productive status in 1907, Bentley in 1908, Maltby in 1910, Yorkshire Main at New Edlington and Bullcroft both in 1911, Askern in 1912, Barnburgh in 1914, Rossington in 1915 and Armthorpe in the 1920s. This was a period during which the hopes of the

[34] See also Bryan E. Coates, 'The Geography of the Industrialisation and Urbanisation of South Yorkshire, 18th Century to 20th Century' in Sidney Pollard and Colin Holmes (eds.), *Economic and Social History of South Yorkshire* (Sheffield, 1976), pp. 14–27.

[35] H. S. Jevons, *The British Coal Trade* (1915, reprinted Newton Abbot 1969 with an introduction by B. F. Duckham), p. 65.

coal-owner were raised by record production figures, just before the collapse of the home and export markets after the First World War.

Just as in the case of the new Dukeries coalfield, the South Yorkshire pits which were sunk in the first quarter of the twentieth century were often accompanied by colliery villages, built by the coal companies, which were seen as models in which the principles of town planning could be exercised:

> It may be hoped that we have left behind the day of the back to back house and the common privy: the industrial colonies erected in the new coalfields—in Kent and South Yorkshire, for example—will bear comparison with some of the more ambitious garden cities, and their owners, while they have had their disillusionments, have certainly gained something by securing and retaining a desirable class of workmen.[36]

Jevons also noted this development:

> A very pleasing feature in some of the mining areas is the awakening of public opinion in favour of more attractive houses and pleasanter surroundings. The Garden City ideal seems to have permeated the community, and a revolt is taking place against building ugly and monotonous cottages in crowded rows.[37]

South Yorkshire, like the Dukeries, was seen at the time as an example of 'Progress' and a portent of the shape of things to come. Here too, areas previously untouched by heavy industry were transformed. Patrick Abercrombie described the village of Armthorpe in his report on the Doncaster Regional Planning Scheme of 1922: 'Until recently a small rural village, with the advent of the new colliery and the housing development in connection with the same, its character has changed and is changing.' Askern, 'originally a Spa noted for its mineral waters and baths, has in recent years entirely changed in character.'[38]

As examples of town planning, some of the new Yorkshire colliery villages did not meet with Abercrombie's full approval. At Armthorpe, 'full advantage has been taken of the new village site', but at Askern the plan of the colliery village was criticized as being too much of a gridiron type, and at New

[36] *Colliery Guardian*, 21 Feb. 1919, p. 432. [37] Jevons, p. 653.
[38] P. Abercrombie and T. H. Johnson, *Doncaster Regional Planning Scheme* (Liverpool, 1922), p. 72.

Edlington the village was 'constructed much too near the shaft of the Yorkshire Main pit' and 'leaves nearly everything to be desired in its planning and the way the work has been carried out, reminding one of some of the early efforts at housing in the South Wales coalfield.' At Barnburgh, 'the Barnburgh Colliery Company's Housing Scheme might have been linked up to the old village, thus forming a community of interest rather than two isolated villages, each with its own centre.'[39]

As in the Dukeries, these colliery villages were isolated, remote from the facilities available in 'town' for shopping, entertainment, and female employment. It is clear too that they did not prove to be the attractive communities the town planners had hoped for. On 11 April 1919 the *Colliery Guardian* quoted from the *Sheffield Telegraph*:

> Within recent years model colliery villages have been built by certain colliery companies. They were certainly model villages by appearance when built, but it is somewhat of a revelation to go round these villages today. In many cases what were neat boundary fences have been pulled up for firewood or utilised for making rabbit hutches, pigeon cotes, etc. The land reserved for the cottage gardens has in many cases been absolutely neglected and allowed to grow wild.

Some of the villages were praised by Abercrombie. Woodlands colliery village, planned on an entirely new site to accommodate workers at Brodsworth Colliery, and designed by Percy Houfton of Chesterfield, was grouped around a magnificent green of trees and regarded by Abercrombie as a model of what new communities should be like. Nevertheless, the social and political institutions of the miners in the new South Yorkshire villages left much to be desired compared with traditional and long-established coalfields. B. F. Duckham wrote in his introductory note to Jevons's *British Coal Trade*:

> His comments on the various districts are full of insight. In the older fields such as Scotland or Durham or Northumberland, division of labour was usually very advanced. Despite wretched housing (especially in Scotland), and other disruptions to family life, such as double and treble shift working, miners had often more self-respect than those in

[39] See above, ch. 3. It was something of a pious hope to believe that a community of interest between new and old villages could be inculcated even when they were situated close together.

> the newer areas and had learnt to assume a more constructive role in the amelioration of their own physical and spiritual environment. Friendly societies, co-operative societies and medical clubs were well supported in addition to the local trade union lodge. In various parts of the Midlands and South Yorkshire, Jevons found a somewhat rougher, less literate type of miner with, on the whole, cruder forms of relaxation.[40]

Despite the optimism of men like Abercrombie who thought that the problems of society could be solved by rational planning within the existing economic system, the new communities in South Yorkshire were often raw, uninspiring villages.

However, there were significant differences between the South Yorkshire colliery villages and the Dukeries. 'Non-political' unionism never made a breakthrough in Yorkshire. The Labour party did organize successfully for local contests in the villages. In 1928, for example, eight out of nine Labour candidates were elected to the Parish Council at Rossington, a village only four miles from Harworth.[41] In the County Council election of the same year, Rossington was placed in the same ward as the old market town of Tickhill. The Labour candidate, a Rossington miner, was elected by 2,530 to 1,550 for the Conservative-Independent, and Rossington miners invaded the old town of Tickhill singing the Red Flag.[42] The social and political quiescence of the Dukeries does not seem to have been shared by new South Yorkshire communities like Armthorpe. The reason for this is not entirely clear. The tradition of greater political and industrial militancy of Yorkshire must have been carried over into the new pits, even though many of them were composed of company housing. In general the Yorkshire villages were less remote than those in Sherwood Forest. Armthorpe, for example, was situated on the outskirts of Doncaster. Perhaps this made it harder for the employers to impose an effective hegemony in the community, although one suspects too that the Yorkshire miners did retain a significantly greater belief in their potential for resistance.

The coalfield whose development most closely coincided

[40] H. S. Jevons, *Coal Trade*, pp. x–xi.
[41] *Doncaster Gazette*, 9 Mar. 1928.
[42] Ibid., 4 Apr. 1928.

in time with that of the Dukeries was Kent. R. E. Goffee's work on the Kent coalfield of the 1920s and 1930s suggests that expanding fields did suffer a considerable degree of dislocation and geographical mobility which contrasts with Bulmer's ideal typification and with the conclusions of such studies as *Coal is Our Life*. Certainly the turnover of labour and union membership in the Kent coal industry was at least as high as that in the Dukeries.[43]

Table 38. *Kent Miners Association: membership and turnover*[a]

	members at years' start	left during year	joined during year
1917	810	371	466
1918	905	357	597
1919	1,145	—	—
1920	1,775	370	600
1921	2,005	449	145
1922	1,701	571	427
1923	1,557	282	325
1924	1,600	230	270
1925	1,640	460	325
1926	1,505	825	525
1927	1,205	516	1,560
1928	2,249	1,040	1,825
1929	3,034	868	1,633
1930	3,799	882	1,580
1931	4,497	570	1,175
1932	5,102	951	675
1933	4,826	551	1,575
1934	5,850	810	1,260
1935	6,300	325	855
1936	6,830	960	960
1937	6,830	749	675
1938	6,756	1,398	525
1939	5,883	698	549

[a] W. Johnson, 'The development of the Kent coalfield, 1896-1946', Kent University Ph.D. thesis 1972, p. 366.

[43] See above, ch. 2.

That the level of turnover shown in Table 38 both represented and caused an atmosphere of confusion and 'cosmopolitan' disarray is clear from other sources, such as the following reminiscences, which demonstrate some common features of new coalfields:

When you couldn't afford to go to the dogs or the pictures you just watched the nearest fight. The Geordies would fight the Scots, and the Welsh would fight the Yorkshiremen and then they'd all change round and fight another lot . . . the Kent coalfield in the twenties was like the Klondyke in gold rush days, and the town of Aylesham, which sprang up for the miners around Snowdown Colliery, was a boom town. There was a turnover of 100 men a week at the start as men 'came full of hope' and 'left in despair'.[44]

Needless to say, all the miners who came to Kent fit into the category of 'long-distance migrants', as there was no established coalfield within 100 miles. There are stories of men hitching or walking down from Lancashire with only the vaguest notions of where Kent was, pushing their belongings in a soap-box.[45] Often these men had been unemployed for months or years, and the strain of working in Kent's deep, hot mines proved too much for them.[46] There was of course no Spencer union in Kent, and there was a general shortage of labour, perhaps due to the isolation and the working conditions, which put the miners in a much stronger position than their counterparts in Nottinghamshire. The companies could not afford to employ the threat of dismissal for the slightest offence when they could not be sure of recruiting miners to replace those they might sack. This, and the fact that the Kent companies could not pick and choose their employees, meant that the Kent coalfield imported Communism along with immigrant miners, and the KMA became a relatively militant union.

The major difference, however, between the new Kent coalfield of the 1920s and 1930s and the Dukeries field which was being developed at the same time was that the Kent field was always at the margin of the British industry.

[44] David Bean, 'Kent and its Miners', *Coal Quarterly*, 1, 8 (1963), 8–10.

[45] R. E. Goffee, 'The Butty System and the Kent Coalfield', *Bulletin of the Society for the Study of Labour History*, 34 (1977), 50.

[46] Gina Harkell, 'The Migration of Mining Families to the Kent Coalfield between the Wars', *Oral History*, (1978), 99.

The existence of a concealed coalfield under southern England, far from any existing centre of mining, had been predicted by the geologist Godwen-Austen as early as 1856, and borings had taken place from 1886 to 1890. But the high investment required for sinking to the deep coal layers was not found until after the First World War, by which time the demand for coal was already dropping. With its poor quality coal suitable only for coking and not for household use, the Kent coalfield could never maintain more than four pits in all.

For a short while, it was thought that a dramatic industrialization would change the face of the Kent countryside:

> The story of the development of the Kentish coalfield, and an account of the vast possibilities of its future, are of particular interest to the present time. The existence in that county of a great field with a vast store of mineral wealth has been amply proved; but as yet it is in its earliest stage of development, for there are only four collieries from which coal is actually being raised and placed upon the market. Several other collieries are in the course of sinking and many more are projected; new railways have been built and others commenced, and eventually some hundreds of miles of railways will be wanted to tap the coalfield. Great docks will also be needed for shipping coal abroad, and to London. A number of ironworks are sure to be started on a large scale and other new industries depending upon cheap coal will arise; and all these surprising new industries will mean the growth in Kent of a hundred villages and many large towns. A strange transformation of the 'Garden of England'.[47]

Like the Dukeries developments, the Kent coalfield attracted the interest of town planners, such as Patrick Abercrombie, whose report on the *East Kent Regional Planning Scheme* studied the effect upon the existing countryside of colliery enterprise and its attendant developments, and tried to produce a plan to facilitate the economic development of the area while safeguarding the beauty of the countryside and the health of its inhabitants.[48] Abercrombie expected that eighteen pits each with 2,500 workers would be sunk in the Kent coalfield, bringing a total increase in population of 150,000 and requiring 28,000 new houses.[49]

[47] Jevons, *Coal Trade*, p. 153.

[48] P. Abercrombie and J. Archibald, *East Kent Regional Planning Scheme* (2 vols., Liverpool and Canterbury, 1925–28).

[49] Abercrombie and Archibald, vol. 1, p. 69.

In fact hardly any of this came to pass, but Abercrombie did design the colliery village of Aylesham, together with John Archibald, the surveyor to Eastry Rural District Council. A model village intended to comprise 1,200 houses and to cost £600,000, Aylesham was to have a long main boulevard starting from the railway station, with a large market square halfway along it. Besides the 552 houses which were built by the coal company (Pearson, Dorman Long), the village comprised a Church of England hall, a Baptist chapel hall, a Roman Catholic church, a Glyn Vivian Mission Hall, a public house, shops, and a temporary school.[50]

Physically the Kent colliery villages were very similar to those in Nottinghamshire dating from the same period: raw, regimented, planned, uniform, isolated communities set in a very attractive rural scene. There were other similarities too. Many miners and their families did not like the isolation of the Kent mining villages, the hostility expressed by the locals, and the lack of neighbourliness they found compared with the North of England. This affected the female immigrants at least as much as the men, as Gina Harkell has shown. One of her interviewees put it this way:

My husband had a lot of friends, but they're not what you call friends, but acquaintances and that. They're social and they speak to you and that you know, but they're not ones for running in each other's houses. I mean to say they speak to you over the garden fence and they're friendly and that, but you don't go in *gossiping* in one another's houses and that you know. The southerners are not so friendly as the northerners, I don't think so . . . up north you could go into your friend's house, and as soon as you went in—'I'm mashin', 'ave a cuppa tea', that's what I missed.[51]

In Kent too the regional differences among the immigrant miners led to tensions, and the butty system of subcontracting imported with the Midlands men caused a great deal of trouble:

For the first three or four years the Welsh stuck to the Welsh, the Derbyshire stuck to the Derbyshire and the Geordies stuck to the

[50] Abercrombie and Archibald, vol. 2, pp. 50–2.

[51] Harkell, *Oral History* (1978), p. 108. This is reminiscent of complaints at Ollerton that women couldn't leave washing on the clothes-line without it being stolen (Butterley H3 224, E. Healey memo, 20 Dec. 1939).

Geordies. If they went into a pub they weren't friendly—there was more trouble than anything else. Everybody used to fight each other over nothing many a time, but mostly over this butty system. A man who wasn't used to the butty system knew he wasn't getting a fair deal so if he was lucky to get hard drunk on a Saturday night, he used to go looking for these fellas that drew all the money and pick a quarrel with them. At Snowdown station I've seen scores of fights there, and it was always buttymen was involved.[52]

There are many features common to both the Kent and Dukeries coalfields. Despite the fact that Kent proved to be an unprofitable coalfield which never lived up to expectations, where the management could not completely curb any discontent, both inter-war developments do suggest that any typification of mining communities as characterized by low geographical mobility and a high degree of social stability must be adjusted to take into account the disruptive effects of a high turnover of labour, itself from varying origins, and the difficulties of building a 'community spirit'.

In addition to the opening up of entirely new coalfields such as those of Kent, South Yorkshire, and the Dukeries, some development of individual mines and seams took place in established coalfields in the 1920s. Let us consider the problems of migration and social dislocation which affected one such development, that of the Markham collieries in North Derbyshire.[53] After the First World War, the favourable market conditions prompted the Staveley Coal and Iron Company to sink two new shafts at Markham Main which increased coal production from 30,000 tons a month in 1924 to over 50,000 tons by December 1929, and employment rose from 1,363 to 1,832 men.[54] Where did the new recruits come from?

Initially miners for the new developments were largely recruited from outside the area. In the early months of 1924 a campaign to secure additional coal getters for the No. 2 pit resulted in the signing of 241 miners from Nottingham, Dudley, Tamworth, Sheffield and Eckington. However within a few months nearly 60% of the total returned home

[52] Harkell, p. 102.

[53] See P. A. Turner, 'Colliery Development in the Inter-war Period, the Opening of the Markham Collieries, Derbyshire, between 1924 and 1930', *Derbyshire Archaeological Journal*, 98 (1978), 83–6.

[54] Turner, 83; Markham Colliery Tonnage and Employment Books, Derbyshire CRO.

including 58% of those from Tamworth, 84% of those from Dudley and all of the Sheffield and Nottingham intake.[55] In contrast, three quarters of those recruited from nearby Eckington remained. The likelihood of problems when dealing with a migrant labour force was therefore indicated from the beginning.[56]

As colliery owner C. P. Markham put it in his evidence to the Samuel Commission,

> I have tried to import labour from other districts into my own district, but this is not a success at some places. The men will not go to work in the older pits, and where the seam is thin. The local conditions are entirely different from what they are accustomed to, and they rarely stay long.[57]

When further labour was needed to man the no. 2 pit in 1925, the company was again forced to recruit from far afield, as there were relatively few unemployed miners in Derbyshire.[58] This process

> was facilitated by the connections the colliery manager had in the North-East, being born in Durham himself. At his initiative a recruiting campaign was started via the local Durham labour exchanges . . . in the second week of February 1926, 117 unemployed miners, chiefly from the West Auckland area, migrated to Chesterfield.[59]

The Durham miners were attracted by the prospects of secure employment at a pit as yet only in the process of sinking, and by Derbyshire's high wages, second only to Nottinghamshire at this time,[60] but they found working conditions very different from what they had been used to,[61] and also met with serious social problems. One local reporter wrote that 'they have left their hearts behind them in Durham',[62] and certainly most of the immigrant miners had left their families behind them when they came down to Derbyshire.

By the end of February 1926, nearly all the Durham men had returned home. It is reasonable to suppose that these men had suffered a considerable social disorientation in their

[55] Royal Commission on the Coal Industry 1925, p. 603.

[56] Turner, p. 83.

[57] Royal Commission on the Coal Industry 1925, Minutes of Evidence, p. 541.

[58] According to the *Ministry of Labour Gazette*, Mar. 1926, p. 91, the unemployment rate in Derbyshire was 1 per cent compared with a national average of 10 per cent.

[59] Turner, p. 83.

[60] MFGB Council Minutes, July 1925.

[61] *Derbyshire Times*, 20 Feb. 1926.

[62] Ibid., 27 Feb. 1926.

migration, for as the company put it, in effect they 'preferred the dole to employment'. After 1926 the Staveley Company employed few men recruited from Durham or from the ranks of the unemployed. Gradually the proportion of 'long-distance' migrants was reduced, to 21 per cent in 1927, 10 per cent in 1929, and 6 per cent in 1930.[63] Many of the migrants tended to congregate in the same mining villages, notably Poolsbrook and Duckmanton, where 200 houses had been erected by the Industrial Housing Association in 1923.[64] These were tied cottages, dependent upon employment at the Staveley pits.

It is clear that the problems of migration apply to the opening of new seams and shafts as well as to the establishment of whole new coalfields. As R. C. Taylor put it: 'The extractive nature of the industry has involved a continual dependence on migration, so that in any one coalfield there are often miners who have been born and worked elsewhere.'[65] The difficulties arising from the opening of new pits may therefore be seen to have a wider relevance than at first sight, for even within established coalfields like Derbyshire new development took place in the inter-war period.

It might have been expected that contemporary developments such as those in Kent and at Markham Main might cause similar problems to those found in the Dukeries in the 1920s and 1930s, and that the process of exploiting the newly discovered coal resources would be approached in a similar manner. But perhaps subsequent changes in the coal mining industry in Britain also demonstrate the same features. If so, then some conclusions may be drawn as to the universal nature of the difficulties and failures of the inter-war years.

The most recent pit to be opened in Nottinghamshire, in 1964, is Cotgrave, the first pit in the county south of the Trent. Cotgrave's notoriety as a 'problem area' for vandalism, crime, and discontent led the Social Services Department of Rushcliffe Council to commission a report which investigated the causes of the village's difficulties. It is possible that some

[63] Turner, *Derbyshire Archaeological Journal* (1978), p. 85.
[64] PRO POWE 16/37.
[65] R. C. Taylor, 'Migration and Motivation' in J. A. Jackson (ed.), *Migration* (Cambridge, 1969), p. 104.

of Cotgrave's ill fame was exaggerated and undeserved, because of the fact that the miners had 'invaded' a district which had recently become popular amongst middle-class commuters to Nottingham, who saw their investment and their home lives threatened. But it is nevertheless clear that genuine social problems did exist, many of which will be recognized by those familiar with the new Dukeries field of the 1920s. The Rushcliffe Report stated that the Social Services felt that 'they were becoming entangled within a morass of social ills within the region.'[66]

First, there were the problems stemming from the immigration of miners to sink and work the new pit. Men came not only from pits closing in the older coalfield of western Nottinghamshire and Derbyshire, such as Radford and Wollaton near Nottingham and the area around Ilkeston,[67] but from much further afield. Immigrants found difficulties in making friends except with members of their own regional group. The transfer of men from Northumberland and Durham

> would appear to represent some of the worst mistakes made at Cotgrave . . . these men from the well-established areas seemed particularly disillusioned on arrival, and a feeling that a false picture had been given prevailed. As people were allowed to choose their own house from those available, small subcommunities grew up, Geordies around Flagholme and the Scots, Welsh and Irish grouped in Whitelands, Little Meadow and Woodview. A certain amount of rivalry occurred which did not help community spirit.[68]

As J. H. Nicholson pointed out in *New Communities in Britain*, 'when newcomers meet for the first time groups from elsewhere, in a setting without established traditions, there is bound to be tension. This may arise even when there is a common occupational background'.[69] In the pubs of Cotgrave, arguments and fights often took place in the early

[66] Anon. (Social Services Department, Rushcliffe District Council), *Cotgrave* (1977), (henceforth 'Rushcliffe'.)

[67] L. M. E. Mason, 'Industrial Development and the Structure of Rural Communities', Nottingham University M.Sc. thesis (Department of Agriculture) 1966, p. 131.

[68] Rushcliffe.

[69] J. H. Nicholson, 'New Communities in Britain', National Council of Social Service', 1961.

years of the village over apparently trivial regional differences such as different methods of scoring at darts or whether greyhounds (a North Eastern tradition) or Alsatians (Nottinghamshire) should be kept. The Nottinghamshire men tended to blame the 'rough, tough Geordies' for stealing, violence, and 'dirty habits', and to use them as scapegoats to be held responsible for the generally poor reputation of Cotgrave.[70]

Another phenomenon, which has also been noted in the Dukeries coalfield of forty years earlier, was the high turnover of people from the new estate, partly by self-removal and partly by eviction. In the first twenty-two weeks of 1965 101 men left Cotgrave Colliery.[71]

> Many of these families came from areas where family ties were traditionally very strong, and a move to Cotgrave obviously seriously disrupted family life . . . it is felt that many families returned home because of this homesickness and in the very least it caused a great deal of strife both mental and physical.[72]

One indication of this was that many children played truant only on Fridays to enable the family to spend a long weekend at their old home. Another cause of the high turnover of labour is also reminiscent of the problems that beset Ollerton in the 1930s:

> Families came to Cotgrave because the mining industry offered employment for the men. Little consideration was given to the fact that there was very little local employment for the women. This, combined with the few social and recreational facilities in the village, often caused women to be much more dissatisfied with living in Cotgrave than the men and is often given as one reason for the high turnover rate of miners in the early days of the colliery.[73]

Besides the fact that many of its first residents left the village in the 1960s, those who remained recorded an extremely high rate of juvenile crime, marital strife, affinity for alcohol, over-use of the National Health Service and bad tenantry.[74] This was not because Cotgrave was financially underprivileged. On the contrary, a low proportion of children were entitled to free school meals. Of the 1,373 households, 699 possessed

[70] Mason, p. 133.
[71] Ibid., p. 138.
[72] Rushcliffe.
[73] Ibid.
[74] Nottinghamshire Community Project Foundation, Cotgrave Report, 1978.

one car and ninety-nine possessed two or more; and the village was mainly populated by relatively well-paid skilled workers.[75] Cotgrave's problems were social, rather than economic, in origin. For example, the incidence of vandalism and juvenile crime may be partially explained by social factors such as the sudden and enforced break-up of the extended family common in the old mining communities from which the Cotgrave miners had migrated. Children roamed the streets at night whilst their parents indulged in social activities, and the influence of grandparents living in the same neighbourhood was noticeably lacking. The age structure of Cotgrave was certainly affected by the absence of 'family elders'.

Another cause of discontent was concerned with the physical aspect of the village. The layout and planning of Cotgrave may be strongly criticized. While 'a need obviously existed to provide housing quickly and relatively cheaply on land that was available', as it turned out 'the estate does present a somewhat awesome and depressing picture with its confusing road pattern and identical housing'.[76] It is true that Cotgrave does not offer much variety. Between 1963 and 1969, 1,048 dwellings were built. Of the male working population of 1,488 in 1971, 1,358 were employed at the pit. Seventy-nine per cent of the miners lived within three miles of the colliery. In Cotgrave itself, only 18 per cent of the housing was owner-occupied, 15 per cent was local authority; and 65 per cent private, rented accommodation, mainly NCB houses, of course.[77]

In addition to this picture of an unrelievedly mining community with its attendant distortions and peculiarities, many amenities and services arrived only after the population did: 'Another bone of contention was that when the estate was first colonised there were not even adopted roads and that the most basic of facilities were a long time in catching up.'[78] The Miners Welfare was not opened until 1966; some observers felt that if it had been operating in the earliest days of the village the fragmentation into regional groups may have been diminished.[79] Shops were inadequate and overcrowded.

[75] Rushcliffe. [76] Ibid. [77] Ibid.
[78] Ibid. [79] Mason, 'Rural Communities', p. 141.

Communications were poor—Cotgrave is not situated on a main road or railway, and the canal is now disused. The lack of a provision of a secondary school is still a major issue in Cotgrave, and school bus accidents as children were taken to the comprehensive school in Radcliffe on Trent led to a 'parents' strike' in September 1978 when children were deliberately withheld from school and a march organized to lobby the County Education Committee in Nottingham.

Rather like the Dukeries villages in the 1920s and 1930s, Cotgrave found itself within the jurisdiction of a local authority previously entirely rural in character. There was certainly a feeling that Cotgrave's problems were neglected by the Rushcliffe Council. After the 1976 municipal elections, for example, the three Labour councillors from Cotgrave were faced by a phalanx of fifty-three Conservatives. But it is likely that this problem was subsumed in the general feeling of hostility caused by the arrival of mining in South Nottinghamshire: 'A certain amount of animosity was immediately established between the older and newer sections of the community. The older community resented the invasion of their peaceful lives by their own conception of a miner'.[80] As Nicholson put it in *New Communities in Britain*, 'When new communities are grafted onto old, the village provides a focus for social activities but is still felt to be "different". The proximity of an established community near, but outside the new estate, may well lead to hostility which is hard to overcome'. Perhaps more because of the resentment of the commuters than of the agriculturally-based inhabitants of 'old' Cotgrave, the miners' estate came to be labelled as a problem area—a label which has stuck.

The new mining village of Cotgrave has suffered dislocation due to rapid labour circulation, the diverse origins of the immigrants, monotonous 'standard' housing, and poor facilities for the provision of education, entertainment and work for female labour. The fact that most of Cotgrave's troubles had been anticipated in the Dukeries coalfield forty years earlier does suggest that despite obvious differences such as the end of private coal-company control, of the

[80] Rushcliffe.

Spencer union, and of the power of the landed aristocracy, and despite the reduction of the feeling of isolation by widespread car ownership, new mining communities do still share common problems. Lessons remain to be learnt. The social implications of the next proposed coalfield, that of the Vale of Belvoir on the Notts.-Leicestershire border are still being debated and disputed. At least, it seems that if Belvoir is exploited, no new villages of a predominantly mining character will be built, but that workers' housing will be added to the existing towns of Grantham and Melton Mowbray to produce mixed, established communities. Sixty-nine per cent of the residents of South Notts. mining villages said they would prefer such an arrangement in a 1957 survey published by the *Journal of the Town Planning Institute.*[81]

The plans to develop the Vale of Belvoir coalfield produced a determined and organized campaign in opposition, which comprised not only conservationists, middle class commuters hoping to protect their property values, and the Duke of Rutland whose castle overlooks the vale, but local shopkeepers and agricultural workers as well. The pages of the *Belvoir Coal News*[82] and the press releases of the Vale of Belvoir Protection Group inveighed against the noise, ugliness, dirt, traffic, subsidence, and health hazards which would accompany the exploitation of a new coalfield. As in the Dukeries in the 1920s, and at Cotgrave in the 1960s, the fear of social change which might be occasioned by the arrival of the miners underlies these reasons for objection.

Parallels may also be sought beyond the British coal industry. Some problems may be found to be common to all mineral extraction industries, and to many countries. There has always been a close, symbiotic relationship between mining and associated population settlements. The communities brought about by the sinking of a colliery are in all countries closely orientated towards the function of raising the coal profitability.[83] It is therefore not surprising that in

[81] Donald Tomkinson, 'East Midland Mining Communities', *Journal of the Town Planning Institute*, 43 (1957), p. 86.

[82] *Belvoir Coal News*, 1-4 (Sept. 1976–Nov. 1977).

[83] Jones, Hull *Occasional Papers*, 1969, p. 3.

the twentieth century, when the multiplicity of small mining concerns has mainly given way to the large units with the necessary resources to sink a deep mine, there may be found in various countries large, compact villages of single-family dwellings, usually of stereotyped plan. As well as the pit villages of England there are the *corons* of France and the *Reihenhaus-Siedlungen* of the German coalfields. These are often cast in the form of planned, functional communities 'at the pit gates'—the *cité-minière* of France, the *zwijnwijk* of Flemish Belgium, the *Bergarbeiterkolonie* of Germany, and the garden village of England.[84]

In many ways too, the company villages of the Dukeries exhibit marked similarities to developments which were taking place in the United States in the inter-war period. Raymond E. Murphy described the village of Maybeury in southern West Virginia in the 1930s: 'Maybeury is a typical coal-mining village, built by the Pocahantas Fuel Company for its miners and consisting of rows of frame houses, monotonously similar in details of construction, and each enclosed by an open board fence.'[85] The company's interest in the village extended far beyond the mere provision of housing:

> The true center of Maybeury, like that of other company villages, is not the post office but the company commissary, a frame building of much the same material and construction as the houses, but much larger. It is the social and business center of the community. Here the men are paid and here is the company store where most of the trading is done, while the porch offers a gathering place for miners after working hours or during temporary layoffs.

Both white and coloured men worked at the mine: 'Though the colored families are segregated, and though their children go to separate schools, their houses and school buildings are just as good as those of the white people.' The valley was entirely non-union, a condition common to much of the Appalachian coalfield.

In *Power and Powerlessness: Quiescence and Rebellion in an Appalachian Valley*, John Gaventa pointed not only to

[84] M. Sorre, *Les Fondemonts de Géographie humaine* (Paris, 1952) vol. iii, pp. 168–71; G. Schwarz, *Allgemeine Siedlungsgeographie* (Berlin, 1959), pp. 242–53.

[85] R. E. Murphy, 'A Southern West Virginian Mining Community', *Economic Geography*, (1933), p. 54.

such features of company control as tied houses, deductions at the pay office for rents, services, goods from the store, and medical care—all provided by the company—but to the strength of the 'third dimension of power', that inculcation of a belief in the dominance of the employer and the powerlessness of the employee which prevented the latter from beginning to think about fighting the inequalities which resulted in poverty in an area of great mineral wealth. The prevailing ideology of the management in Clear Fork Valley (Tennessee and Kentucky) precluded any criticism of the way in which 85 per cent of the land was owned by an absentee (British) fuel company. Law and order, respectability, religion, and patriotism were called into play against the forces of disorder, especially that brought by 'outside influences': 'one drop of pure Kentucky blood is worth more and is more sacred than an entire river of Communist blood.' Attempts to unionize the Appalachian miners or to stimulate political militancy led by northern liberals or left-wing activists failed. In Gaventa's view this was because of the way the miners had absorbed the values and morality passed down by the companies, just as it might be argued that many inhabitants of the villages of the Dukeries coalfield accepted the dominance of the company over their lives even outside working hours, and accepted the notion of a quiescent 'model village'.

Parallels may also be drawn with other types of mining, such as that of Southern Africa. The compound system had been invented by Cecil Rhodes's company, De Beers, in the 1880s, as African workers were found to be smuggling small and easily hidden diamonds out of mines. By erecting a completely closed compound with fences and towers guarding the perimeter, the companies' profits could better be protected. Besides preventing smuggling, compound systems were soon found to have other advantages. Desertions and consequent labour shortages were minimized. The compounds were run on a 'barracks' principle which emphasized the authority of the company officials who were often uniformed, like the company policemen of Nottinghamshire: 'The provision of uniforms served not only to separate the "police" from the workers, but it also provided a gloss of legitimacy for the violence that was an integral part of the job.'[86] Discipline was a rough business: besides the imposition of fines and the loss of privileges, the whip or *sjambok* came into play often, and mine officials, protected by the judicial system, could expect to escape lightly if they injured or even murdered their workmen.[87] The 'third dimension' of power came into play here too, for the African was always denied the status of a mature man, and condemned to the role either of a savage or a perpetual 'boy'. It is scarcely surprising that the miners tended to despair of the possibility of any kind of action to relieve their plight. In addition, they were disunited. As the Nottinghamshire coal companies segregated miners by regional origins, the African companies were known to play on ethnic differences to divide the workforce, and often provoked fights between tribal groups which they then claimed proved the African's unfitness for independence.[88]

Van Onselen saw the African mining companies' emphasis on sports and recreations as 'defusing class consciousness'. Tribal dances were encouraged because ritual diverted the miners' minds from more serious issues. Since dancers were largely ethnically based, this had the effect of reinforcing divisions among black workers on tribal lines.[89] Compound sports were largely initiated by the manaement in a deliberate effort to defuse potentially dangerous situations. As the Native Commissioner for Belingwe put it in 1920, a programme of organized weekend activities was needed:

> For a moment let us consider what it was that made the British proletariat contented, although working in many cases, in circumstances which were scarcely more conducive to a sustained interest in their labours than are those in which the mine boys work here. It was largely sport—or what the workmen considered sport. For example, the hands old and young in every community were enthusiastic 'supporters' of some local football team whose Saturday afternoon matches furnished a topic of interest for the remainder of the week. Here the labourers' principal recreations are connected with beer and women, leading frequently to the Police Court and the risk of being smitten with one or another of the venereal diseases which are so insidiously sapping the strength of the native population. Those who employ and those who

[86] Charles Van Onselen, *Chibaro* (London, 1976), p. 140.
[87] Ibid., pp. 143–50. [88] Ibid., pp. 158–94. [89] Ibid., pp. 186–9.

control native mine labour should, for a double reason, try to influence the native to change in this respect. Sporting enthusiasm is not the ideal substitute for present conditions, but it would be a step forward, and one, I am sure, not difficult to bring about. The native is intensely imitative, often vain, and always clannish, and all these are qualities which would further 'sport'—a parochial spirit of sport if you like, but one which would forge ties of interest and esprit de corps between the labourer and his workplace. A patch of ground, a set of goal posts and a football would not figure largely in the expenditure of a big mine.[90]

In many ways, of course, the African compounds seem far removed from the British coalfields of the 1920s or 1930s, and indeed the control exercised by the companies in the Dukeries was nowhere near as unsubtle. There were no visible fences and there were no beatings, and the principal threat behind any coercion was that of dismissal and unemployment. Nevertheless, certain important principles may be considered to apply in both cases. The idea of isolated housing being provided for its workers in the vicinity of the mine enabled the employers to keep a constant eye on their employees. As Van Onselen put it, 'employers needed a system which aimed at control of the worker both in *and* outside of his working hours'. Mission education and the established Christian churches were encouraged by the African mining companies and regarded as part of their investment. By inculcating the values of the colonizing people the churches could articulate an ideology supporting the mining industry. The Bishop of Mashonaland implored the African that 'thou shalt keep thy contract'. On the other hand the 'subversive' Ethiopian Watch Tower movement was suppressed.

Migration was of course another feature of the African mineral industry. 'Circulation of labour' took place as migrant workers moved from rural areas to boost their earnings, then returned to their tribal homeland as soon as they could.[91] In some ways this represents a conflict between economic motives—the desire to be able to possess the goods brought by the white man to Africa—and social pressures forcing the

[90] Ibid., p. 191.

[91] J. Clyde Mitchell, 'Labour Circulation in Southern Africa', in Jackson (ed.), *Migration*; *idem*, 'The Causes of Labour Migration', *Bulletin of the Inter-African Labour Institute*, 1959.

worker to fulfil his tribal obligations. In any case it is certain that a high degree of geographical mobility was common to all the African mineral extractive industries.

Despite obvious differences in the form of company control and the type of labour migration in the United States, Africa, and Britain, common problems and solutions did stem from the natural propensity of mines to be situated in remote and isolated areas, often not previously industrialized before the exploitation of their mineral resources. The need for the employer to provide housing near to such mines meant that companies could maintain an unusually complete control over the lives of their employees at and away from work. Often this implied political weakness on the part of the miner, and industrial organization was made difficult not only by the employer's influence but by the high degree of labour turnover inherent in a cyclical extractive industry. This pattern of migration and renewal of the work-force does much to give the history of mining, especially the exploitation of new resources, its unstable and complex character. The new British coalfields of the twentieth century are unusually well-documented examples of phenomena which will recur as long as the earth's natural resources are tapped.

12. The Dukeries Model Villages as New Communities

'Starting from scratch . . .'

The new colliery villages built to house the miners who came to work in the Dukeries between the wars were seen by their builders and owners, the coal companies, as 'model villages'. By this they meant that they were not only a model of attractiveness, with their neat semi-detached houses in broad avenues laid out on logical geometric principles, but also planned communities, conceived and executed by a single authority which retained ownership and control of the whole village as a functional unit. Starting from scratch, the companies had *carte blanche* to construct villages to their own specifications to create the type of community they required. They were very proud of the result, which they saw as a pointer to the future, and a model which might with profit be copied. In many cases this view was successfully passed on to the inhabitants of the new villages. This chapter will consider how the coal companies' conception of a planned village compares with various other attempts to build new communities in modern Britain, what motives lay behind the shape these ideas took, and to what extent common problems faced all those creating these new communities and those who lived in them.

Model villages in the Dukeries

It was almost universally recognized at the time that the housing provided in the Dukeries colliery villages was of a substantially better quality than that in the areas from which the immigrant miners came. The companies publicized their achievement widely, especially the Butterley Company, which scoured technical journals for opportunities to write of its scheme whereby hot water was piped around the village of New Ollerton after having been heated at the pit.[1] B. S. Townroe described Ollerton model village in *The Spectator* in 1926:

One of the most interesting of the new colliery villages is at Ollerton, about a quarter of a mile north-east of the old village, and on the main line between Chesterfield, Lincoln and Grimsby. Here some 1,100 houses are to be built. A special effort has been made to avoid monotony by standardization of types, and as brick is the material used, it has been possible to have ten different types of houses. These vary in size from 946 square feet to 866 square feet. The average cost after the Chamberlain subsidy is about £420. This includes not only the house and foundations, but garden paths, fencing, entrance gates, connexions with main sewer, electric lighting and the installation of a system of central heating and hot water supply.

This hot water system has been developed by the Butterley Colliery Company, who own a number of pits in the district, and is alone worth a visit as one of the most notable experiments in providing an unceasing flow of water practically at boiling point for the use of colliers and their families at all times of the day and night, and at all seasons of the year, at an estimated cost of sixpence a week. Although the Coal Commission strongly recommends the provision of pithead baths, many miners whose homes are not too far away from the pithead prefer to go straight home to wash and change their clothing. At Ollerton the hot water supply is heated by the exhausted steam at the pithead. At week-ends or when the pit is standing, live steam is used. The hot water is circulated to all houses on the estate, and only drops in temperature very slightly after the whole circuit is completed.

Other features of this mining village are the spacious rooms, the number of cupboards, the playgrounds for children and the beauty of the lay-out. All those interested in housing can learn many lessons from this admirable industrial housing scheme.[2]

Other contemporary observers who discussed the new coalfield noted the 'social value of comfort':

New mining villages are rising all over the Dukeries—Blidworth, Warsop, Harworth, Clipstone. The worst canker of mining social life has been eliminated. Here we have none of those huddled shanties, rotten with age and dirt, uninhabitable and uninhabited by thousands upon thousands of our coal getters in every coal district in the kingdom. Here we will have none of the ghastly rows of ill-built, insanitary houses, ringed with iron bands to guard against pit subsidence, that exist in the older Notts field. Here in the open land of the Dukeries the narrow meanness of the owners has been discarded. They have shown a wider outlook—a realisation of the social importance of cleanliness, of the social value of

[1] Butterley H3 223, *Sheffield Independent*, etc.
[2] B. S. Townroe, *The Spectator*, 10 Apr. 1926.

comfort—which marks an almost revolutionary tendency in their class. The new villages are built of substantial houses, ugly enough in their square brick rawness, dear enough in rent, small enough too, but habitable and comfortable. The outside miner looks with envy and discontent on their bathrooms, on their electric light, on their pit-head baths, on their running water. Housing is the first of the great magnets that draw the best blood of the English mining areas to the Dukeries.[3]

The authors of the Butterley Company history contrasted the new mining village of Ollerton with the dead colliery village of Llewellyn's *How Green Was My Valley*. The Samuel Commission noted the difference between the old-style colliery communities of the established coalfields and the model villages of the new Nottinghamshire and South Yorkshire fields:

The housing conditions of colliery workers, like almost all else connected with the industry, show great diversities. They are often very bad—many of the old villages consist of poorly constructed cottages, small and frequently overcrowded, with sanitary arrangements primitive and inadequate, the aspect of the villages being drab and dreary to the last degree. At the other extreme there have been built by some of the large new colliery companies, garden villages, which are well planned, well constructed and well equipped, equal to any in Great Britain.[4]

The reason for this transformation was that in nineteenth-century mine undertakings entrepreneurs had been unwilling to invest substantial sums in workmen's housing because of the likelihood that the coal seams would be worked out within a few decades and because of the prevalence of short-term mineral leases of fifty or sixty years, or even thirty-one years in Scotland, where the housing was notoriously inadequate.

On the other hand, where modern collieries have been established, to work seams that will last a long time, and under leases running for a lengthy period, the proprietors are better prepared to spend very large sums of capital in the erection of good housing accommodation. They not only have the satisfaction of rendering a public service, but they are enabled to draw to their undertaking the best class of workers. At some of the large mines in the South Yorkshire and Notts coalfield, from one third to one half of the total capital provided is being spent

[3] *New Statesman*, 24 Dec. 1927.
[4] PP 1926 xiv, Royal Commission on the Coal Industry, p. 213.

upon houses for the workers, sums of £750,000 to £900,000 being devoted to that purpose by industrial undertakings.[5]

G. C. H. Whitelock described Bircotes colliery village, built for Harworth Colliery, as 'a model of town planning'. A resident architect, Philip N. Brundell, was appointed with a staff of four.[6] According to Whitelock, they carried out their duties with 'vision and foresight', seizing the ideal opportunity to start from scratch in laying out a town of 1,100 houses of a variety of designs, all with hot and cold water and bath, tree-lined roads, wide paved footpaths, and walled gardens. The company also built an Institute with concert hall, billiard room, and card room, and provided twenty-one acres of sports grounds; a church, church hall, and parsonage were constructed by Barber-Walker at their own expense. The company even provided the name of the village: 'the new village was christened Bircotes'.[7]

This demonstration of the complete planning of a village was not unusual. All the colliery companies operating in the lucrative new Nottinghamshire field built villages at the pit gates to ensure the 'necessary stable reserves of manpower'.[8] In general, the companies let tenders for the building of the new villages to local firms, but at Blidworth the work was entrusted to the Industrial Housing Association, who planned towns on what they called 'garden suburb principles'.[9] Sir John Tudor Walters, who wrote the history of the IHA, *The Building of Twelve Thousand Houses*, pointed to the benefits of their dwellings: at least three bedrooms, hot water supply to coal owner Charles Markham's own design, sites reserved for public houses, institutes, shops, and recreation. Tudor Walters also claimed that state subsidies for housing were unnecessary; house-building should be run on a business footing according to the laws of supply and demand, and there were 'obvious dangers in the wholesale ownership of houses by local authorities'. The North Midlands Survey

[5] Ibid., pp. 213-14.

[6] G. C. H. Whitelock, *250 Years in Coal, the History of the Barber-Walker Company 1680-1946* (Derby, 1955), p. 251.

[7] Whitelock, p. 169.

[8] See above, ch. 4.

[9] Sir J. Tudor Walters, *The Building of Twelve Thousand Houses* (London, 1927), pp. 24-31. See Fig. 5.

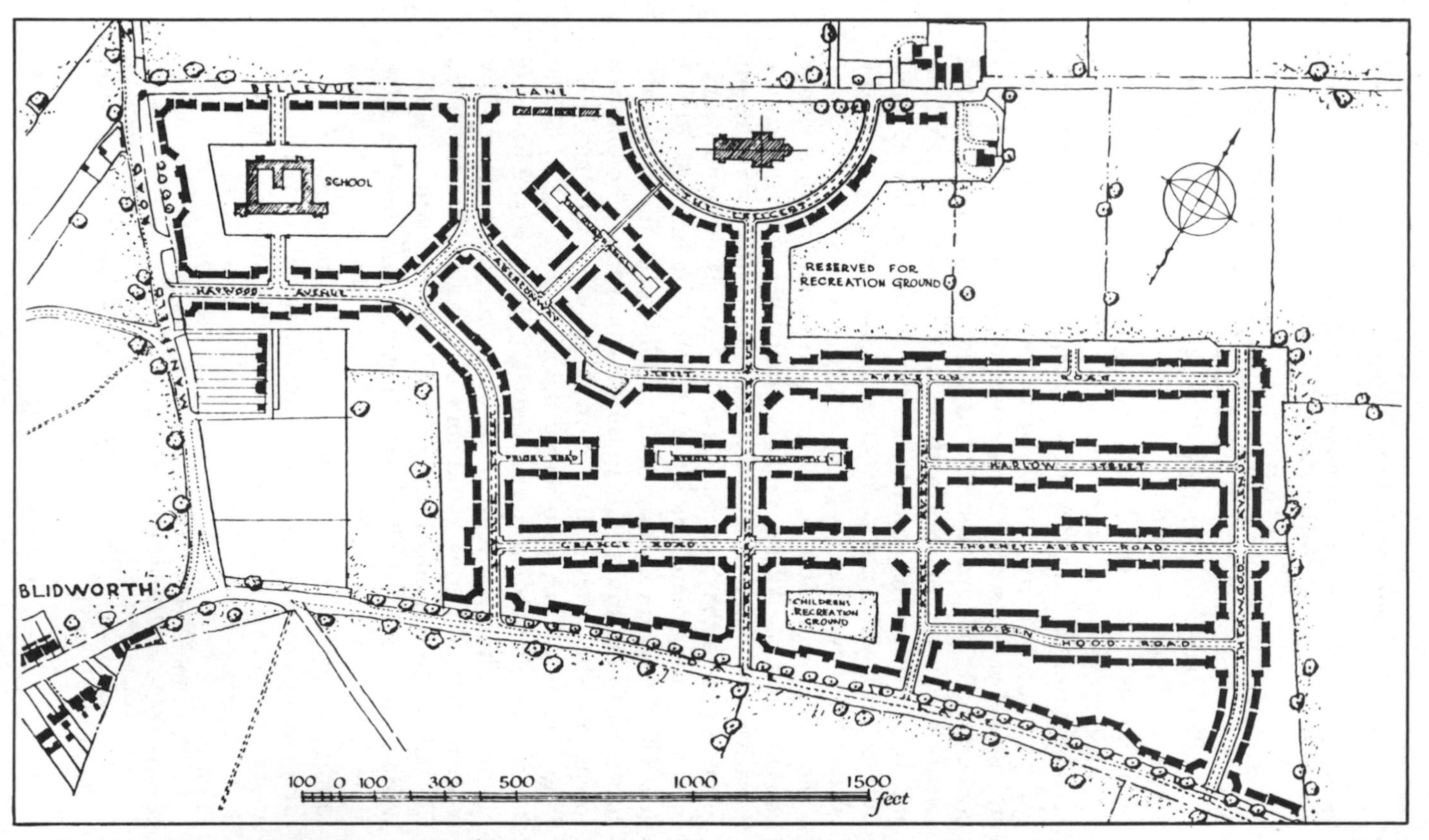

Fig. 5. Blidworth Village Lay-out Plan (From Sir J. Tudor Walters, *The Building of Twelve Thousand Houses*, London 1927)

Report of 1945 praised the housing conditions in the new concealed Notts. coalfield, and singled out the Butterley hot water system. In the Butterley Company's files on Ollerton, the colliery village was always referred to as 'the model village' or as 'New Ollerton'. It was seen as the company's own creation and their own property too. Indeed the model villages were more than simply a device for attracting men to work at the pit; they were designed to influence their behaviour once they were there.

There is no doubt that the building of an isolated colliery village 'at the pit gates' could act as a force for disciplining the workmen. The company's regulations for tenants were designed to maintain cleanliness and order. Pets or animals of any kind were often not allowed in company houses:

FT There was none allowed in the village.
GC Oh no, dogs weren't allowed.
RJW That was part of the housing contract, was it?
FT Oh yes.
RJW And why did they insist on that?
FT Cleanliness, nothing else.
FJW And that ended when? Nationalization?
FT Yes, like a good many more things![10]

Gardens were to be kept tidy, on pain of trouble at the pit. Lawns were compulsorily mowed by the company, and the cost stopped out of the men's wages. Noisy and 'immoral' behaviour was frowned upon.

The values of the company, and their attitudes towards the colliery model villages were shared by many of their employees:

RJW Has the character of the place changed a lot?
JES Yes.
RJW In what way?
JES Deteriorated.
RJW Why has it deteriorated?
JES Well, it was a model village in the old days.
RJW And now you think that all sorts of people have come in and the unity and character of the place has changed?
JES Well, litter was almost unknown in those days.
RJW Is that because people wanted to keep the place tidy themselves? Did the company have people to check to make sure it was tidy?

[10] Interview, George Cocker, Frank Tyndall, Edwinstowe.

JES Well, the company employed people that had been injured in the colliery at the work, you see, and they were employed on light jobs; and all the lawns were mowed once a week.
RJW So the place looked very neat and smart?
JES Oh, by Jove, not half![11]

The miners and their families certainly noticed a stark contrast with the older mining villages from which most of them had come:

RJW Your family came from Tibshelf;[12] didn't they?
DT Yes.
RJW Why did they come to Edwinstowe?
DT Because there was no work. My father worked at Blackwell. Because they were on three days, and three days on the dole, and there was seven days work to be done here. And as I say, we came from a house where there was no lights, we had Aladdin lamps; and no bathroom, no hot and cold, no flush toilet or anything. It was heaven! . . . it was a showplace.
FT It was a show village.[13]

Further examples of the quality of the housing may be found in other Dukeries colliery villages:

CB Well, I think for those days the housing was good. I mean whenever you see programmes of the miners in the old days, you know in sort of back-to-back houses, and I could never understand this, because every colliery house I'd ever lived in had had a bathroom in it and it had had lovely hot water, it had had a garden, and I couldn't understand the grimness that was portrayed in, well, film programmes in those days.
BB They were very well-built houses, very well maintained.
CB I can remember a little bit of damp in one of our cupboards, but apart from that—
BB This house I lived in in Edwinstowe, though, it looked a terrific size, didn't it?[14]

Harworth colliery village exhibited similar attractions:

When I came here, I'd been brought up in large rows of ugly houses, two up and two down, we used to call them, bare back to bare back along row outside toilets, earth lavatories, unhygienic, many times filthy, seeing the cart come to empty 'em, the effluent streaming from

[11] Interview, J. E. Smith, Ollerton.
[12] Tibshelf is situated on the old coalfield of North Derbyshire.
[13] Interview, Dorothy Tyndall, Frank Tyndall, Edwinstowe.
[14] Interview, Barbara Buxton, Forest Town (Edwinstowe) and Cyril Buxton, Forest Town (Ollerton).

the cart . . . and then I came up here. And it struck me that this was probably at that time of the day the most modern village in the world. In the world! And how nice it was, how pleasant, you just walked in and knocked a switch down on the wall and the light came on! What a bloody big change! And lots of people, who were coming here at that time from Lancashire, Derbyshire, all over, they were leaving a very similar environment to what I'd had, and they were overjoyed, and they were writing home do you know, the women folk in particular were writing back and saying 'come and see us!' and you could buy postcards in the shops, ordinary postcards, with scenes of various streets, and they used to go and buy these postcards and 'this is our home!' and mark it with a bit of a cross, you know, and send it, 'come and see us!'[15]

Now we must judge to what extent the coal company villages of the new Nottinghamshire field did live up to the claims that they were 'models' of planning. How do they compare with other attempts at constructing new communities in industrial Britain? What problems did they have in common with garden cities and new towns, new municipal housing estates, other company villages and rapidly growing single industry boom towns? How did the ideas of the creators of the Dukeries mining villages compare with other notions of how a new community should be planned?

Garden cities and new towns

First of all, it would not be correct to claim that the Dukeries mining villages did owe much to the ideals of Ebenezer Howard and the garden city movement, which were current in the 1920s and 1930s. Most of the key aims of Howard and his disciples, and indeed of the government sponsored new towns which followed after the Second World War, were missing from the coal companies' plans. The model colliery villages were not attempts to solve the social problems brought about by the unplanned growth of industrial cities, as was Howard's policy of dispersal of population and factories. They altogether lacked the grandiose conceptions of men such as Lewis Mumford, who wrote:

At the beginning of the twentieth century, two great new inventions took form before our eyes: the aeroplane and the Garden City, both

[15] Interview, Neville Hawkins, Harworth.

harbingers of a new age: the first gave men wings and the second promised him a better dwelling place when he came down to earth.[16]

The New Town Commission of 1946 also demonstrated the self-confidence and faith of the new town movement. In the post-war period of reconstruction, the Commission believed it better to start from nothing, with a clean sheet, than to extend existing communities. This was a policy that was later to be called into question.[17]

There was however no attempt to make the Dukeries mining villages self-contained communities attracting a variety of industries. This was how the garden cities of Letchworth and Welwyn differed from council housing estates and purpose-built private suburbs. Indeed the garden cities were to be 'not a suburb but the antithesis of a suburb',[18] a microcosm of the richness of British society as a whole. Planned residential suburbs like that at Hampstead were known,[19] but Howard's Letchworth, designed by Raymond Unwin and founded in 1903, was the first whole town planned according to a policy of low housing density. There was to be an attempt to ensure a 'social balance' in the garden cities, and middle and working classes were to live together in the same part of the community, in harmony. Unwin and Barry Parker formed a link between Howard's private garden city movement and government interest in the ideals of town planning. During the First World War they worked for the Ministry of Munitions and were responsible for planning the layout of the camps at Gretna Green and Eastrigg, and they later collaborated in designing the municipal housing estate of Wythenshawe, Manchester (1927-41).[20] Parker on his own was responsible for several model colliery villages,[21] but these were far removed in principle from the mixed community of

[16] Introductory Essay by Lewis Mumford to 1946 edition of Ebenezer Howard, *Garden Cities of Tomorrow* (London, 1898, 1902, 1946), p. 29.

[17] See below, pp. 263-6.

[18] Mumford, p. 35.

[19] See the special issue of *Town and Country Planning*, 25 (1957), 275-320 on Hampstead Garden Suburb.

[20] Gordon E. Cherry, *The Evolution of British Town Planning* (Leighton Buzzard, 1974), p. 74; M. G. Day, 'Sir Raymond Unwin and R. Barry Parker: a study and evaluation of their contribution to the development of site planning theory and practice', Manchester University MA thesis 1973, pp. 122-6.

[21] Day, p. 147.

a true garden city. 'As long as social classes exist, all must be equally represented' was also the hope of the New Towns Commission.[22] This was a far cry from the rigidly enforced hierarchy of the Dukeries villages. The planned colliery communities had limited, functional aims, concerned more with maximizing the company's profit than securing 'a better tomorrow'. Social harmony was seen as an aid in securing continuity of production rather than as an end in itself.

Despite this difference in conception, there were problems which both kinds of new community suffered. In both cases the existing inhabitants of the region resented the arrival of new industry and immigrant workers.[23] At Stevenage, for example, Harold Orlans and Lloyd Rodwin have described the opposition to the 'invasion' of Londoners which resulted in higher rates, great changes in the character of the town, a loss of agricultural land, planning restrictions and a loss of local control.[24] Second, there was a tendency in the new towns as in the Dukeries colliery villages to provide housing well in advance of vital services and amenities. As the Reith Committee's Report had put it, 'for those (amenities) which are essential in a new town, building priority must be given, but some will have to wait their turn.'[25]

Unfortunately, progress in provision of these services was slow, and the effects were unhappy, despite Frederic Osborn's pious hopes: 'There are inevitable cultural and service shortages during the first phases. But if there are good houses and reasonable roads and paths for getting about, the new residents will find the pioneering life not only tolerable but uniquely enjoyable.'[26] At Stevenage the first factory did not arrive until January 1953. There were no sewers at Basildon by 1956 and the education service was poor: 'Burdened by the growth at Harlow and elsewhere, Essex County was tardy in meeting its educational commitments.'[27] Harlow was isolated,

[22] New Towns Committee, Final Report, PP 1945-6 xiv, p. 22.

[23] See above, ch. 3.

[24] H. Orlans, *Stevenage* (London, 1952), pp. 160-3; L. Rodwin, *The British New Towns Policy* (Cambridge, 1956).

[25] New Towns Commission, Final Report, PP 1945-6 xiv, pp. 727-809, Cmd. 6876, para. 5.

[26] F. J. Osborn, *Green Belt Cities* (London, 1946), p. 157.

[27] Rodwin, p. 109.

with a poor transport system, as well as inadequate water and sewerage schemes. The New Town Development Corporation Annual Report for 1956 admitted that the schools in Newton Aycliffe were 'grossly overcrowded'. In Basildon in the same year, 'there was no progress . . . in the provision of local health services. The present situation is a constant source of hardship to the local population.' Similarly at Harlow, 'there has been much local dissatisfaction at the delay in the provision of hospital facilities for the town.'[28] At Stevenage in 1957 it was reported that the 'comparative sparseness of its amenities' was 'the most serious threat to the well being of the town'.[29] Another set of problems concerned the clash of authority between the new town development corporations and more local bodies. Conflicting jurisdictions led to chaos; a lack of local control over the development of the new towns bred resentment. The sign at Stevenage railway station was changed to 'Silkingrad' as an expression of local discontent with the control exercised by Lewis Silkin, at the time Minister of Town and Country Planning.[30]

So, although it is possible to castigate the Dukeries coal companies for not aiming to build balanced, democratic communities, we should not underestimate the problems facing those who founded new towns. The British new town experiment has not proved a notable success. In 1956, a critical American observer remarked that

> Some of the experiences to date are, alas, melancholy to contemplate . . . the building of the new towns has disclosed more frustrating limitations and problems than were anticipated by the planners, and has required far more flexible tools, subtle understanding, and calculated foresight than were brought to the task. The weaknesses are evidenced by the assumptions, the experiences and some of the changes in direction and emphasis. To the extent that these weaknesses persist, the programme may fail to achieve its ultimate promise.[31]

As with the Dukeries villages, the problems were a mixture of 'growing pains' and difficulties arising from the concentration of power necessary in a rigorously 'planned' community:

> In these planned towns a bewildering variety of unanticipated problems have cropped up during the early years. The first model town[32] became

[28] New Towns Commission, Annual Report, PP 1957–8 xvi.
[29] Ibid.
[30] Orlans, *Stevenage*, p. 69.
[31] Rodwin, *New Towns Policy*, p. 5.
[32] Stevenage.

an example of bad practice. Towns that started with grass roots support for a long time steadily lost that support. The garden city leaders, the most vigorous exponents of the programme, have been outraged at the treatment of one of their experiments.[33] The resort to a corporate device to achieve administrative flexibility has wound up with incessant protests against the Whitehall straitjacket. The planning mechanism has shown embarrassing failures in forethought. In short, there has been a deplorable amount of fumbling.[34]

Besides the failings that beset the new town movement in practice, the principle behind their foundation has itself come under attack. It is no longer thought that advanced town planning can solve society's problems. The government *Social Survey* of 1947 found evidence that people did not want to live in mixed-class areas, and that such experience did not change attitudes in the way that had been hoped.[35] In *Town and Country Planning* for 1950 Gordon Campleman argued the case for single class neighbourhoods.[36] Peter Collison stressed that it proved beyond the scope of town planning to end the social divisions in British society:

> Although planning policy may do something to modify the degree of (class) segregation it cannot be expected to eliminate it completely nor can it be expected, of itself, to bring about any profound changes in the social structure. If attempts are made to mix the social classes in close proximity, it seems likely that these attempts will be resisted, and as more dwellings become available, increasingly ineffective.[37]

Heraud traced the way in which areas of new towns became associated with particular social groups, despite efforts to produce a mix of all classes on the part of the development corporations.[38] Hudson and Johnson produced a Marxist critique of the new town movement, stressing that although revolutionary in appearance, new towns were always conceived

[33] Welwyn Garden City, Howard's second experiment, founded in 1919, which was designated a new town in May 1948, giving power to a development corporation in place of Howard's private company.

[34] Rodwin, *New Towns Policy*, p. 129.

[35] Bertram Hutchinson, 'Willesden and the New Towns', *The Social Survey* (1947), pp. 40-1.

[36] Gordon Campleman, 'Mixed Class Neighbourhoods; some Sociological Aspects', *Town and Country Planning* (1950), pp. 330-2.

[37] P. Collison, 'Neighbourhood and Class', *Town and Country Planning*, 23 (1955), p. 337.

[38] B. J. Heraud, 'Social Class and the New Towns', *Urban Studies*, 5 (1968), 8-21.

of in terms of protecting the existing structure of society—'a mechanism through which the State attempts to resolve crises in capitalism'. The 'role of the New Towns, their Development Corporations as branches of the State, is social control'.[39]

Whether or not one accepts the stance of these writers, one must certainly agree that the new towns, like the garden cities, have been firmly placed in a tradition of commercialism, their success or failure judged in terms of financial viability. Howard and his followers believed not in a socialist society but in the regulation of the unplanned development of capitalism. His garden cities were owned by private companies. The Reith Committee, on the other hand, adopted the principle of state intervention. It now seems clear that new towns are no panacea for the ills of modern industrial society, as the early idealists and the post-1945 optimists hoped. William Alonso could write in the 1970s:

On the whole, a national policy of settling people in new towns is not likely to succeed, and it would not advance the national welfare if it could be done. The principle flaw in new town proposals lies in an underestimation of the social and economic integration and connectivity of a modern society . . . even if new towns turned out to be wonderful places, this would be almost irrelevant to our present urban problems, and as sirens of Utopia they might distract us from our path.[40]

Perhaps there was not such a glaring gap after all between the garden city and new town movements and the individualist commercialism of the Nottinghamshire coal-owners. Despite their claims, neither form of new community really broke with the traditional manner of organizing society, yet both faced problems in their early years consequent upon the attempt to plan urban development from scratch.

Council housing estates

If the colliery villages lacked the lofty ideals of the new towns, there were less ambitious types of venture in industrial Britain which may also offer close comparisons, if the problems

[39] R. Hudson, M. R. D. Johnson *et al.*, *New Towns in North-East England* (Durham, 1976), pp. 22, 422.
[40] W. Alonso, 'What are new towns for?', *Urban Studies*, 7 (1970), 54.

of young and homogeneous communities in general are to be considered. These include twentieth-century local authority housing estates; nineteenth-century factory villages, usually constructed by a single employer for his workforce; and the 'boom towns' of nineteenth-century industrialization, usually dependent on a single industry.

Such developments show how social and demographic change is an inevitable concomitant of economic transformation. A high level of migration, both short-distance and between various parts of the British Isles, was a response to the changing nature and location of industry and employment in modern Britain. In the 1930s a number of studies were undertaken which attempted to analyse the movement of labour in Britain from areas of declining industry such as Lancashire and South Wales to the new-industry towns of South Eastern England.[41] Some dramatic results were uncovered. At Dagenham, for example, 28.7 per cent of the workers had migrated from elsewhere in the year before July 1932. At the steel town of Corby in 1937, 49.2 per cent of the workforce had arrived in the last twelve months, two-thirds of them from Scotland. A general movement from North to South resulted in the proportion of the insured population living south of a line drawn from Portsmouth to the Wash increasing from 24 per cent in 1923 to 28 per cent in 1937. Needless to say, many housing developments were needed to cater for such a shift in population. What kind of problems were faced by the new estates and communities?

Let us consider the municipal council estates which were being erected at the same time as the new Notts. mining villages were being built, in the 1920s and 1930s, and also the post-Second-World-War council estates. Studies of these have

[41] J. Jewkes and H. Campion, 'The Mobility of Labour in the Cotton Industry', *Economic Journal*, 38 (1928), 135-7; Brinley Thomas, 'The Movement of Labour into South Eastern England', *Economica*, NS 1 (1934), 220-41; *idem*, 'The Influx of Labour into the Midlands 1920-37', *Economica*, NS 5 (1938), 410-34; *idem*, 'The Influx of Labour into London and the South-East 1920-36', *Economica*, NS 4 (1937), 323-6; R. S. Walshaw, 'The Time Lag in the Recent Migration Movement within Great Britain', *Sociological Review*, 30 (1938), 278-87; A. D. K. Owen, 'The Social Consequences of Industrial Transference', *Sociological Review*, 29 (1937), 331-54; M. Daly, 'Reply to A. D. K. Owen', *Sociological Review*, 30 (1938), 236-61; A. D. K. Owen, 'Rejoinder', *Sociological Review*, 30 (1938), 414-20.

revealed many problems found also in the Dukeries. First, the high rate of turnover of population as residents of the new estates rapidly left, for one reason or another. This turnover was concealed by the general trend towards the new estates. In the five years from the completion in 1951 of the Essex estate of 'Greenleigh', which was mainly intended for the resettlement of London East Enders, to March 1956, 26 per cent of the tenants left.[42] An earlier Essex estate, that originally known as Becontree but now usually called Dagenham, was constructed between 1921 and 1935. With its 27,000 houses, it grew from almost nothing to become the largest housing estate in the world.

Table 39. *Population of Dagenham estate*[a]

Mar. 1922	2,086	1928	57,820
1923	11,837	1929	77,455
1924	14,564	1930	82,869
1925	19,089	1931	91,519
1926	26,241	1932	103,328
1927	40,071		

[a] Terence Young, *Becontree and Dagenham* (London, 1934), pp. 38, 48.

In its early years, the removal rate from Dagenham was between 7 per cent and 17 per cent per annum, compared with an average of 4 per cent for other LCC properties—'people were coming and going as fast as you could see.'[43] Another council estate developed contemporaneously with the Dukeries coalfield was that of Watling in Hendon, North London. Ruth Durant's study of Watling claimed that the vast turnover of inhabitants disjointed the growth of any kind of permanent community, describing the estate as 'a vast hotel without a roof'.[44] Leo Kuper's survey of a new working class council estate in Coventry, reported that one third of the families left within a period of four years.[45]

[42] Michael Young and Peter Willmott, *Family and Kinship in East London* (London, 1957), p. 164.

[43] Peter Willmott, *The Evolution of a Community* (London, 1963), p. 20.

[44] Ruth Durant, *Watling* (London, 1939), p. 119.

[45] L. Kuper (ed.), *Living in Towns* (Birmingham, 1953), p. 139.

Kathleen G. Pickett and David K. Boulton found that in Kirkby, Lancashire, 'One third of Kirkby residents moved into the town against their will. The same proportion expressed a desire to move away.'[46] Nor was this unpopularity of new housing estates in the Liverpool Merseyside area uncommon: at Norris Green, a council estate built in Liverpool between 1926 and 1929, 30 per cent of the original inhabitants had left within two years and 50 per cent within the first ten years.[47] Jevons and Madge described the new inter-war Bristol Corporation estates thus: 'The population of the new estates was constantly increasing before the war as new houses sprang up; new faces were appearing every week. At the same time many people were moving away again.'[48] The effects of this level of population circulation on attempts to create a unified community spirit are clear. Hilda Jennings wrote in her study of Barton Hill in Bristol, that after three and a half years, 'exactly half of the 90 families which had moved in from older districts had moved again or were awaiting removal. This state of affairs did not conduce to a state of belonging or a desire to become involved in estate activities.'[49] Ruth Durant concluded that as far as Watling was concerned, 'the constant turnover of its population was the greatest single handicap to its developing as a community'.[50] In post-war Glasgow council estates, 'in the early days of the scheme as many as 40% of the tenants had their names on the list of applicants for transfer'.[51] Even when economic and housing conditions made it difficult for the discontented actually to leave an estate, the attitudes of residents could still be surveyed. One third of those living on a Sheffield estate in 1952 wished to leave,[52] as did one third of respondents in the Cheshire overspill town of Winsford in the 1960s.[53]

[46] K. G. Pickett and D. K. Boulton, *Migration and Social Adjustment* (Liverpool, 1974), p. 39.

[47] N. Williams, 'Problems of Population and Education in the New Housing Estates', Liverpool Ph.D. thesis 1938.

[48] Rosamond Jevons and John Madge, *Housing Estates* (Bristol, 1946), p. 64.

[49] Hilda Jennings, *Societies in the Making* (London, 1962), p. 220.

[50] Durant, *Watling*, p. 119.

[51] T. Brennan, *Reshaping a City* (Glasgow, 1959), p. 57.

[52] G. D. Mitchell, T. Lupton, M. W. Hodges, and C. S. Smith, *Neighbourhood and Community* (Liverpool, 1954), p. 90.

[53] H. B. Rodgers and D. T. Herbert, *Overspill in Winsford* (Keele, 1965).

Why did so many people 'vote with their feet' on the subject of living in new housing estates, just as they did in the new Nottinghamshire coalfield between the wars? One disadvantage that both possessed was a lack of services and amenities consequent upon the failure of the authority responsible for the new community to provide these essentials in advance of the arrival of the population. As Brennan remarked in the case of new post-war council estates in the city of Glasgow, 'The pressure of need and the balance of political power in Glasgow was such that the policy adopted was one of "houses first and as rapidly as possible". Everything else was left for later consideration, and, of course, difficulties have arisen.'[54] The adverse consequences of the rapid development of new local authority housing were by no means limited to Glasgow. For example, the disruption of education in the early years of a community is frequently to be found.[55] Terence Young's Pilgrim Trust survey of Dagenham found that 'the Dagenham village Infants' School during 1924–5 doubled its numbers and became full. Marsh Green school . . . added another 100 to its roll and had an excess of 60.'[56] At Watling, Ruth Durant described the state of secondary education in the 1930s as very unhappy.[57] The diverse origins of the immigrants did not help: at Mossdene, a new estate in Bristol, one new school drew its pupils from thirty former schools.[58] In Glasgow in the 1950s, 'large numbers of children had to be transported by special buses between the estate and older parts of the city, and some children went through the whole of their school life being unable to attend a local school'.[59]

Transport was also given little attention in general. Dagenham was poorly served by buses, and the train service to London was almost non-existent until 1932, a very important factor in the boredom and isolation of the estate in an age before the mass ownership of motor cars. On a new Liverpool council estate constructed in the 1940s, 'in the early days of

[54] Brennan, *Reshaping a City*, p. 40.
[55] See above, ch. 7.
[56] Young, *Becontree and Dagenham*, pp. 58–9.
[57] Durant, *Watling*, pp. 62–9.
[58] Jennings, *Societies in the Making*, p. 122.
[59] Brennan, *Reshaping a City*, p. 53.

the estate, the inadequacy of transport to the city centre eight miles away was a constant source of complaint.'[60] This made it all the more serious that a lack of provision of entertainment and leisure services was also a feature of the new estates. There were no new pubs built in Dagenham before 1928, and no library by 1934, when the population of the estate exceeded 100,000 people. The Bristol estate of Sea Mills did not get a public library until 1934.[61] Brennan wrote of the Pollok estate in 1959: 'a temporary public library was opened in 1949, but there is still no permanent building nor any prospect of one.'[62] Voluntary organizations were handicapped in the early years of the new estates: 'These found it difficult to work because of the lack of accommodation. Without schools, there were no school halls available for letting.'[63] Few facilities for young people were provided at Watling in the 1930s, leading Durant to doubt whether 'leisure can supplant work as a means of social fusion?'[64] Pickett and Boulton summed up Kirkby in its early years as 'like Siberia' because of its lack of leisure facilities.[65] Clubs, associations, and other organizations which might have bred community spirit were notably absent. Hilda Jennings painted a picture of council-house tenants faced with an omnipotent bureaucracy, unable to act for themselves or participate in local decision making.[66]

This was certainly not unconnected with the one-class nature of most of the council housing estates. Dagenham, for example, represents an extreme of social homogeneity: over 92 per cent of its inhabitants were defined as working class in Moser and Scott's study of 'British Towns' in the 1950s.[67] Despite the advantage of social harmony, this did mean that the traditionally middle class institutions of societies and clubs were missing. That the new estates were seen as working class was one reason for the high rate of migration from them. Families with social aspirations felt it

[60] Mitchell, Lupton, Hodges and Smith, *Neighbourhood and Community*, p. 19.

[61] Jevons and Madge, *Housing Estates*, p. 67.

[62] Brennan, *Reshaping a City*, p. 54.

[63] Ibid.

[64] Durant, *Watling*, p. 120.

[65] Pickett and Boulton, *Migration and Adjustment*, p. 61.

[66] Jennings, *Societies in the Making*, p. 216.

[67] C. A. Moser and Wolf Scott, *British Towns* (Edinburgh, 1962), p. 34.

necessary to move away in order to achieve a higher status. Jevons and Madge found that 28 per cent of those leaving Bristol estates in a survey sample taken between the wars did so because of their disapproval of the 'invasion' of slum families, and a further 15 per cent left in order to buy their own house. The very people most likely to act as 'community leaders' were those most inclined to be discontented with life on the new estates.[68]

Council estates have had a 'bad press'. In the years between the wars, for example, there took place the most remarkable example of class segregation and distrust in twentieth-century England. When a council estate was built next to a private housing development in the North Oxford district of Cutteslowe in the early 1930s, the Urban Housing Company which had built the private estate erected seven-foot high walls topped with iron spikes across the roads linking the two estates. Clive Saxton, the head of the housing company, said that this was necessary because of the influx of 'slum families' into the council estate.[69] A survey by Peter Collison found that in the 1950s 73 per cent of the private tenants described themselves as middle class and 84 per cent of the council tenants thought of themselves as working class.[70] Despite much local protest, the Cutteslowe walls remained in position until 1959—a period of some 25 years. The view of council estates as criminal and violent working-class ghettoes did nothing to foster their community spirit, and made it even more difficult to obtain a satisfactory blend of occupations and classes in new, planned units.

The lack of formal organizations to unite the inhabitants of the housing estates would not have been so serious on its own, but, as Jennings put it, informal associations were cut out just as much as formal associations were. Shops were more remote due to the absence of 'corner shops' on the estates, so there were fewer opportunities for contact with the neighbours. In *Family and Kinship in East London*, Willmott and Young argued that the extended family, so much the keystone of working-class life in a traditional

[68] Jevons and Madge, p. 67.
[69] P. Collison, *The Cutteslowe Walls* (Oxford, 1963), p. 77.
[70] Collison, pp. 38-9.

community like Bethnal Green, was seriously eroded by the migration of the younger members of East End families to low density housing estates in Essex. One is reminded of the social problems in the new colliery village of Cotgrave in Nottinghamshire, exacerbated by the absence of grandparents to look after the children. J. M. Mogey found the same phenomenon in his comparison of the old working class neighbourhood of St. Ebbe's in Oxford with the post-war council estate on the edge of the city at Barton.[71] This had serious social consequences. Hilda Jennings summed up thus:

> In the absence of humanising and personalising influences of place operating through services and established kinship and neighbour groups, the isolated one generation family on the estate tended either to remain aloof with one or two 'special friends' or to seek a better-class neighbourhood.[72]

It is true that attempts were made in the Dukeries coalfield to maintain contacts with relatives in Derbyshire or Durham, and that homesickness for the more intimate community of the wider families of the older coal-mining villages led both to discontent and to outward migration.[73]

The differing origins of the immigrants to a newly founded town or village, so noticeable in Nottinghamshire, was also found in new housing estates. The regional variations did not have to be dramatic, or based on differences in social class. As Willmott and Young found, most of the immigrants to 'Greenleigh' in Essex came from the East End of London, but quoted one such immigrant as saying 'we all come from the slums, not Park Lane, but they don't mix'. The authors added:

> Such a vast common origin might be enough to bind a group of Cockneys in the Western Desert; Essex is much too near for that. When all are from London, no-one is from London; they are from one of the many districts into which the city is divided. What is then emphasised is far more their difference than their sameness. The native of Bethnal Green feels himself different from the native of Stepney or Hackney.[74]

[71] J. M. Mogey, *Family and Neighbourhood, Two Studies in Oxford* (Oxford, 1956).

[72] Jennings, *Societies in the Making*, p. 221.

[73] See above, p. 50.

[74] Young and Willmott, *Family and Kinship*, p. 125.

More problems were caused by the age structure of these new communities. In Bristol council housing estates between the wars, it has been estimated that half the population was aged nineteen or under.[75] At Kirkby in the 1950s and 1960s the concentration of children at their most destructive stage gave 'the illusion of a total youth population of delinquents'.[76] Moreover, problems were 'stored up' by the odd age structure. When the huge numbers of children at Dagenham grew up, married and started families, housing was not available for them; and so Dagenham, the overspill town for London, had to pass its own overspill on to Canvey Island. One husband remarked bitterly: 'It's a ridiculous situation. The LCC puts its overflow into Dagenham. Dagenham puts its overflow, of necessity, out to Canvey Island. Where they go from Canvey Island in the next generation, God knows. We can only assume they'll put them on rafts and set them adrift.'[77]

Besides these specific problems, further generalized indicators of discontent can be cited. The highest incidence of mental illness in the country in the 1950s was found to be on an LCC housing estate in Hertfordshire.[78] The crime rate at Kirkby has long been notorious; Pickett and Boulton described it as 'a vandal's paradise and a criminologist's challenge'.[79] The lack of neighbourliness and the destruction of the extended family in the new housing estates led Ruth Durant to reflect on the position of their women: 'The loneliness of the people here in the first few months after their removal to Watling is extreme. The women are mostly affected by that desperate loneliness.' In sum,

> The inhabitants of twentieth century housing estates have less feeling of belonging to a local society and are less friendly to their neighbours in general . . . in some ways this would appear to be conforming to the way of life already established in a number of middle-class districts, in which privacy is thought of either as a good in itself, or as a necessary adjustment to social status.[80]

[75] Jevons and Madge, p. 27.

[76] Pickett and Boulton, *Migration and Adjustment*, p. 38.

[77] Willmott, *Evolution*, p. 38.

[78] F. M. Martin, J. H. F. Brotherton and S. P. W. Charre, 'Incidence of Neurosis on a New Housing Estate', *British Journal of Preventative and Social Medicine* (1957).

[79] Pickett and Boulton, p. 37.

[80] Jennings, p. 216.

New housing estates may therefore be seen as one of the causes of decline in 'traditional proletarian' working class consciousness, and the rise of privatization, which so disrupts a view of a unified working class in the second half of the twentieth century in Britain.

Factory villages

The second type of new community which may be compared with the mining villages of the new Notts coalfield is the nineteenth-century factory village. By this is not meant the experiment of George Cadbury at Bournville, where the housing was not intended to be retained in company ownership or even confined to employees at the associated factory.[81] Bournville fits more into the Utopian tradition of community-builders like Ebenezer Howard. Factory villages are sometimes defined as communities built by a manufacturer expressly for the purpose of housing his own workforce near the place of employment, and regarded as part of the investment entrusted to the enterprise.

Perhaps the most complete example of such a factory village, although there were many in nineteenth-century industrial Britain, was that of Saltaire, a prime example of mid-Victorian individualist self-confidence. The Bradford textile magnate Titus Salt attempted in middle life a grand design of replacing his existing mills with one huge factory outside the town, built between 1851 and 1853.[82] For the next twenty years the process of constructing a community to house the workers at the factory continued. The prospectus for the town of Saltaire showed that it did espouse ideals of town planning. The *Manchester Guardian* reported that

the architects are expressly enjoined to use every precaution to prevent the pollution of the air by smoke, or the water by want of sewerage, or

[81] For Bournville, see Bournville Village Trust, *When We Build Again* (1941); *Sixty Years of Planning, the Bournville Experiment* (1960); *Bournville Housing* (1922, 1926 edns.).

[82] For the history of Saltaire and the details below, see William Ashworth. *The Genesis of British Town Planning* (London, 1954); R. Balgarnie, *Sir Titus Salt, Baronet: His Life and its Lessons* (1877); A. Holroyd, *Saltaire and its Founder* (2nd edn. 1871); R. K. Dewhirst, 'Saltaire', *Town Planning Review*, 31 (1960/1), 135-44.

other impurity . . . wide streets, spacious squares, with gardens attached, ground for recreation, a large dining hall and kitchens, baths and wash-houses, a covered market, schools and a church; each combining every improvement that modern art and science have brought to light, are ordered to be proceeded with by that gentleman who has originated this undertaking.[83]

As with the Dukeries mining villages, the ownership stressed the physical attractions of their plans, plans for a closely regimented village.

Housing was the first priority in Saltaire; parlour houses cost £200 to build, while ordinary workmen's cottages cost £120. Again, the employers bore the cost of the 'public' buildings in Saltaire. In 1859 a Congregational Church was completed at Salt's own expense; nine years later a Wesleyan Chapel was added by public subscription on a site donated by the owner. Temporary accommodation for elementary schooling was provided from the beginning, and in 1868 permanent buildings were erected at a cost of £7,000. Salt was as proud of his drainage system as the Butterley Company were to be of Ollerton's hot water scheme. An accident infirmary and dispensary were erected at Saltaire, as were baths and washhouses (1863) and forty-five almshouses. There was a factory canteen, and in 1871 a club and institute was provided, with games, art, and library rooms, although all public houses were banned from Saltaire 'with true non-conformist zeal'.[84] With unconscious irony, the *Birmingham Post* summed up Saltaire:

It is the result of thought and design, the realisation of a good idea . . . it has at least shown what can be done towards breaking down the barrier which has existed between the sympathies of the labourer and the employer. The founder of Saltaire has taught us that there are noble duties which the capitalist can perform, and that in discharging them he may elevate himself to a glorious position without interfering with the prosperity of his business.[85]

Yet whatever the benefits of life in Saltaire, it cannot be seen as the realization of a progressive idea. It was a monument to the dominance of the ego of the Victorian factory owner,

[83] *Manchester Guardian*, 21 Sept. 1853.
[84] Ashworth, *Genesis of Town Planning*, p. 128.
[85] A. Holroyd, *Saltaire and its Founder* (London, 1871), pp. 31-2.

and was far removed from ideas of equality and democracy later to become more current in British thinking about the manner in which society should be organized. Paternalism was not to be the way forward.

Port Sunlight, founded by W. H. Lever in 1888, was also closely connected to an industrial enterprise, and indeed the management of the new community was entrusted to a special department of the firm—'the tenants for many years could air their complaints only to the firm's estates manager'.[86] Almost all the houses in Port Sunlight were occupied by the company's employees. Subsidized by Lever's other departments to provide lower rents and free maintenance, Port Sunlight nevertheless proved to be no drag on the company's prosperity but rather a sound investment, maintaining a peaceable workforce at the factory gates, as subject to the company's influence at home as at the factory.

As Sidney Pollard has shown, there are countless other examples of factory villages in nineteenth-century England.[87] Almost all of these involved the provision of a new community as part of the entrepreneur's investment and as an aid to prosperity. Robert Owen's much vaunted New Lanark was no exception. The 'model' community there could attract labour to a remote area and retain it even when Glasgow could offer higher wages.[88] Discipline was instilled into the workforce, and drunkenness, which could so curtail productivity, might be curbed. Symptoms and causes of criminal behaviour were tackled, and patrols kept checks on household cleanliness and sobriety. In this way an 'efficient, trustworthy and stable workforce might be built up'.[89] There was nothing socialist about this aspect of Owenism—New Lanark was an efficient capitalist venture, and its paternalist treatment of its employees differed little from earlier examples such as the factories of the Arkwrights at Cromford in Derbyshire and

[86] Ashworth, p. 134.

[87] S. Pollard, 'The Factory Village in the Industrial Revolution', *English Historical Review*, 79 (1964), 513-31.

[88] A. J. Robertson, 'Robert Owen, Cotton Spinner: New Lanark 1800-25' in Pollard and Salt (eds.), *Robert Owen, Prophet of the Poor* (London, 1971), p.152.

[89] Ibid.

the Strutts at Belper.[90] Later in the nineteenth century there were to be many other cases of an employer founding a new village to serve his works. Certain general principles can be isolated concerning this policy, which Pollard sums up thus: 'Here were whole townships under the social and economic control of the industrialist, their whole raison d'être his quest for profit, their politics and laws in his pocket, their quality of life under his whim, their ultimate aims in his image.'[91] These communities could be dominated by 'benevolent' employers as in the case of New Lanark, or perhaps not—as the Hammonds put it, the factory villages were 'not so much towns as barracks . . . these people were not the citizens of this or that town, but hands of this or that master'.[92]

Usually the factory villages were established in order to house labour in a remote area in order to tie it more closely to the place of work: 'The decision to establish a large works in the open country was taken to be synonymous with the need for new housing.'[93] Pit villages were a clear example of this, in the nineteenth century as in the twentieth. Although housing was the first priority in the factory villages, other buildings could fulfil a useful function; elementary schools were not usually built by the employers, except in Scotland, but Sunday schools were. One possible role for them is that mentioned by Pollard: 'Sunday Schools had a far more important role to play, being designed largely to inculcate current middle class morals and obedience.' Hannah More thought that education should be used in the Mendips 'to train up the lower classes to habits of industry and virtue',[94] and the Quaker Land Company's schools were responsible for the absence of 'Chartism, Radicalism, and every other abomination' from the villages.[95] The early Mechanics' Institutes were intended to make the working

[90] R. S. Fitton and A. P. Wadsworth, *The Strutts and the Arkwrights* (Manchester, 1958), p. 98.

[91] Pollard, p. 513. See also J. D. Marshall, 'Colonisation as a Factor in the Planting of Towns in NW England' in H. J. Dyos (ed.), *The Study of Urban History* (London, 1968), pp. 215-30.

[92] J. L. and B. Hammond, *The Town Labourer* (London, 1917), pp. 39-40.

[93] Pollard, p. 517.

[94] M. G. Jones, *The Charity School Movement* (Cambridge, 1938), pp. 158-9.

[95] A. Raistrick, *Two Centuries of Industrial Welfare* (London, 1938), pp. 63 ff.

classes more amenable and to teach 'respect for property'.[96] Dissenting employers would build chapels, like Salt did, and it is in this tradition of general education and training that we can place the expense set aside by the Notts. coal-owners for the new churches built in their villages in the 1920s and 1930s.[97]

Besides the role of the churches and the schools, there were other ways in which the factory villages could be used as behavioural regulators. The employers were 'forced to take account of the worker's behaviour outside working hours, of his family, the likelihood of his migration, and of his attitude to the factory system as a whole, often called by contemporaries his "moral outlook".'[98] The economic power of the owners in these villages is clear enough. It can be seen in the use of truck, and the resource of arbitrary dismissal, which would mean removal from the community in a factory village of tied housing. Pollard concludes that factory villages did little to prevent the manufacturers from regarding their workers as 'hands' without brains, to be kept occupied lest they be led into mischief. After all, as Smelser has pointed out, great though the outward difference was between the flogging masters and the model community builders, 'from the standpoint of the control of labour, both types of factory management display a concern with the enforcement of discipline.'[99] The manufacturer who found himself in charge of a factory village, like Kirk Boott of the Massachusetts Lowell Company, had to be 'its town planner, its architect, its engineer, its agent in charge of production, and the leading citizen of the new community.'[100] What is more, in Lowell 'the architecture everywhere proved the integration of the ideological and industrial plan'.[101]

In nineteenth-century Lancashire, paternalist employers created a cultural, political, and social hegemony that transcended mere 'social control', converting power relationships into moral ones,[102] and elevating the factory system to an

[96] Pollard, p. 527. [97] See above, ch. 8. [98] Pollard, p. 521.

[99] N. J. Smelser, *Social Change in the Industrial Revolution* (London, 1959), p. 105.

[100] Hannah Josephson, *The Golden Threads* (New York, 1949), p. 39.

[101] P. and P. Goodman, *Communitas* (New York, 1947), p. 62.

[102] Patrick Joyce, *Work, Society and Politics* (Brighton, 1980), pp. 91–2.

accepted way of life as central and unquestionable as the family structure itself.[103] Tory bonhomie and Liberal discipline were the varied styles of a pervading influence which was all the more successful for its rejection of condescension and its concealment of dependence.[104]

The Dukeries colliery villages must be seen in a predominantly nineteenth-century tradition of paternalist social control, a control to which isolated mining communities were especially susceptible. As Pollard put it, 'colliers lived in isolated and easily tyrannised communities'. Control over the lives of their employees outside working hours was seen as an essential element in protecting the investment involved in each industrial enterprise. With the growth of mass communications and the decline of the 'local society' in the twentieth century, different methods and organs of control would have to be found. The Dukeries mining villages represent one of the last major examples of the company village in Britain, inviting many comparisons with the factory villages of a previous era.

Boom towns

The factory villages of the nineteenth century were necessarily small; larger communities could not be so easily controlled by a single employer, and the function of a company village would be vitiated by growth. Nevertheless, another nineteenth-century phenomenon in Britain also illustrates some of the early problems of the Dukeries villages. These were the one-industry boom towns of the Industrial Revolution. Just as was the case in the new Notts. coalfield, they were seen as a wonder, and a pointer to the future full of promise. Gladstone compared Middlesbrough to 'an infant Hercules, but an Hercules just the same'.[105]

Just like the new towns, the new housing estates, and the new coalfield communities, they had to cope with a rapid growth of population. This meant that there was a constant feeling of living on a building site: 'Seafaring men . . . used to

[103] Ibid., p. 186. [104] Ibid., p. xviii.
[105] Quoted in Asa Briggs, *Victorian Cities* (London, 1963), p. 247.

Table 40. *Population of Middlesbrough*[a]

1801	25	1871	39,563
1831	154	1881	55,943
1841	5,463	1891	75,532
1851	7,431	1901	91,302
1861	19,416		

[a] Source: A. Briggs, *Victorian Cities* (London, 1963), p. 253.

Table 41. *Population of Barrow in Furness*[a]

1859	800	1871	18,911
1861	3,135	1881	47,259
1864	8,176	1891	57,712

[a] J. D. Marshall, *Furness and the Industrial Revolution* (Barrow, 1958), pp. 227, 407.

talk to one another from their own doors through speaking trumpets, same as they did at sea, because they could not travel between the two houses without sinking up to their calves of their legs in the mire.'[106] Mud was a common factor in the early days of various kinds of new communities. Certainly it was noticeable at Ollerton:

> When I first came by bus from Mansfield, train to Mansfield, I remember, and then a bus out to Ollerton, I asked them to put me off at the picture house. They did, and when I alighted off the bus I went over the ankles into soft, sandy mud and I said 'well, where's the picture house?' and they said 'behind that hedge', and there it was![107]

and at Dagenham: 'You went right up to your ankles in mud and slime in the winter. There were awful muddy holes in the road, and the women used to pick up bricks from the building site as they went past and drop them in the holes in the road to give us stepping stones for our feet on the way back.' The impression that Middlesbrough was a 'frontier town' was reinforced by the variety of origin amongst the inhabitants. In 1871, 50.1 per cent of the population of Middlesbrough had migrated from South Wales, Scotland, Durham, or Ireland. Briggs summed up Middlesbrough: 'Rapid growth,

[106] Briggs, p. 251.
[107] Interview, A. E. Corke, Ollerton.

the heterogeneous composition of its population, and the preponderance of the male sex, recall features generally credited only to the towns of the American West.' In Barrow in 1871, males over twenty outnumbered females by 6,170 to 3,777.

The diverse origins of the inhabitants of these new communities could lead to clannishness. The rapidly expanding settlement of Merthyr Tydfil in the 1840s forms one example. The Commission of Inquiry into the State of Education in Wales reported that 'there it was remarked that there was a constant immigration of workmen who lived together clannishly, the Pembrokeshire men in one quarter, the Carmarthenshire men in another, and so on.'[108] At Coatbridge in Scotland too, migration consequent upon industrialization brought social disruption:

> When the manufacture of malleable iron began there in 1839 most of the skilled workmen needed were brought from England and Wales, especially from Staffordshire. These men were at first paid highly and set themselves apart from the rest of the population, with whom they were frequently brawling. . . . at Coatbridge, the process[109] was delayed by the influx of more and more Irish, some of them deliberately maintained as strikebreakers, from whom the original coalminers of the district long held aloof.[110]

Contacts were maintained with the donating areas, and a considerable degree of chain migration took place—the early inhabitants of Barrow were known to hang around the railway station, awaiting news from their relatives at 'home'. All this adds up to a picture of a typical boom town, which Briggs could call a 'turbulent urban frontier' in the case of Middlesbrough, where one might find that 'a large part of the inhabitants was given up to intemperance'.[111]

Like many 'boom towns', most of the British examples of the nineteenth century were dominated by one industry, and to some extent by one employer. In the case of Crewe,

> the company had made itself fully responsible for the running of the town. It paid the salary of a clergyman who superintended the day schools, and gave a house and surgery to a doctor who attended all the

[108] *Report of Commission of Inquiry into the State of Education in Wales* (1847), Part I, Appendix, p. 304.

[109] i.e. of integration.

[110] Ashworth, *Genesis of Town Planning*, p. 32.

[111] Briggs, p. 251.

railway workers' families in return for regular fixed weekly subscriptions. The steam pump at the locomotive works was also used to provide an adequate water supply for the town, and it used the coal-tar produced there, mixed with gravel and ashes from the railway workshops, to provide an asphalt surface for the footpaths of the town streets.[112]

Often the employer was directly responsible for the planning and building of the town. James Ramsden of the Furness Railway Company was in charge of the planning of modern Barrow, employing 'free use of the T-square' to produce a rectangular street pattern, with broad thoroughfares that 'anticipated the arrival of the motor-vehicle'.[113] J. D. Porteous described the layout of Goole by the Aire and Calder Canal Company's Chief Engineer George Leather as a product and a symbol of company control, with its vigorous, uniform architecture, and each street and quarter designed for a specific purpose, with a rigid segregation of social classes.[114] The 'new town' of Middlesbrough was laid out on a strictly symmetrical plan in order to promote 'some uniformity and respectability in the houses to be built'.[115] There was a central square with the unimaginatively named North Street, South Street, West Street, and East Street leading away at right angles to each other. All in all, Porteous's typification of company towns could be applied without much distortion to the Dukeries:

> The high degree of control of workers in company towns is reflected in major physical and social features which include a rigid ground plan, architectural uniformity, dominance of the townscape by company institutions, deliberate allocation of social classes to housing of distinct quality types, and strong ethnic and class segregation.[116]

Local government, such as there was, was also dominated by the companies. Henry Bolckow, the ironmaster, was the first mayor, the first president of the Chamber of Commerce

[112] Ashworth, p. 28. See also W. H. Chaloner, *The Social and Economic Development of Crewe 1780–1923* (Manchester, 1950).

[113] J. D. Marshall, *Furness and the Industrial Revolution* (Barrow, 1958), p. 233.

[114] J. D. Porteous, 'The Company Town of Goole', University of Hull, *Occasional Papers in Geography*, 12 (1969); *idem*, *Canal Ports: the Urban Achievement of the Canal Age* (London, 1977), pp. 144–9.

[115] Briggs, *Victorian Cities*, pp. 250–1.

[116] J. D. Porteous, 'Social Class in Atacama Company Towns', *Annals, Association of American Geographers*, 64 (1974), 409.

and the first MP for the town of Middlesbrough, returned unopposed in 1868. He also gave the town its first public park and new schools for 900 children. In Barrow, resident director James Ramsden was empowered on 9 February 1867 to represent the Furness Railway Company 'at any parish meetings at which the company may be interested'. He was the first mayor of Barrow (1867-72), a director of most of the employing companies in the town, a JP, a founder alderman, a member of the Board of Guardians and the School Board, the Colonel commanding the local rifle volunteers, president of the Cricket Club and of the Sanitary Association, and vice-commodore of the Yacht Club. A statue was put up to him in 1872. The first town council was composed overwhelmingly of professional men: ten out of seventeen members worked either for the Railway Company or the Haematite Company. The Barrow oligarchy supported the Liberal MP for North-West Lancashire, the Marquess of Hartington, eldest son of the seventh Duke of Devonshire (Barrow had no separate parliamentary representation until 1885). There was no Conservative organization in the town at all until 1873, and J. D. Marshall could write: 'Political divisions were to have no real meaning in any case; there was only one "party", and it was ruled from the railway offices and the steelworks.'[117] The council met not in the newly built town hall but in the Railway Company's boardroom (the railway company had built the town hall, market hall, and police station in any case).

However, the *de facto* control of the companies was rarely marked by efficient local government:

> During this period of breakneck urbanisation unprecedented in the history of the region, Barrow lacked any form of effective local government . . . in 1864, when the new Furness steel town and port had over 8,000 inhabitants, it lacked any legally controlled authority having power to make a simple bye-law for sanitary improvement. It lacked baths, workhouses, a suitable poor relief system, a hospital and even an adequate supply of medical practitioners; a service for the disposal of filth and nightsoil, firefighting equipment, a local magistracy, a burial place, an adequate postal system, and even numbering of houses.[118]

[117] Marshall, *Furness*, p. 299.
[118] Ibid., p. 288.

Education was similarly disrupted. In Barrow the company spent £600 on a school and reading room in 1858 and £365 for the enlargement of the school in 1861, but there was no school board until 1872; still in 1873 only 2,000 out of the 5,000 children eligible were receiving any formal schooling. At Middlesbrough, schools had been in existence since the 1830s, but it had proved difficult 'to prevail or compel the attendance of the children'.[119] In Barrow there was no Poor Law relieving officer within six and a half miles until a new Poor Law district was created in 1871.

Churches were popular with the employers: 'The planning was concerned with spiritual welfare as well as with material wellbeing.'[120] Indeed the Haematite Steel Company of Barrow-in-Furness gave £4,500 for a church on 15 November 1865, but donated only £500 towards the town's much needed hospital, which was opened that year. Earlier, 'the energetic Ramsden had caused the church of St. George to be erected on Rabbit Hill, Barrow Village, in 1859-60, and accordingly a district chapelry was created.'[121]

As with the new council estates, as with the other nineteenth-century boom towns, and as with the Dukeries mining villages, Barrow took on the aspect of a frontier community. Houses were erected so quickly in the 1850s and 1860s that 'Barrow, with its streets unmade or roughly filled with furnace slag, looked more and more like a vast builder's yard.'[122] There were great differences, of course, as well as similarities between these boom towns and new mining villages. As they grew, it became increasingly diffcult for one company to control towns like Middlesbrough and Barrow, and their turbulence contrasts with the tranquillity of more fully controlled communities. In Barrow in the 1860s there was an active branch of a trade union, the Amalgamated Society of Engineers. There were strikes, even a rattening. The period of the boom over, these towns were to settle down into a period of greater maturity and stability, and enjoy a more healthy balance of authorities. But traces of their early history remained. It was still possible for Lady Bell to write of Middlesbrough in 1907 that it was a community

[119] Briggs, *Victorian Cities*, p. 252.
[120] Ibid., p. 251.
[121] Marshall, *Furness*, p. 281.
[122] Ibid., 286.

of a 'pre-ordained, inevitable kind, the members of which must live near their work', in those cheap houses near the factory gates in 'little brown streets'.[123]

Predominantly single-company towns did continue to grow in the twentieth century. Brian Didsbury argued that in the case of Northwich,

in the 1890s Brunner Mond gradually assumed a position of dominance as the major source of employment in the district, and Northwich, their original base, became, in effect, a company town. From their position of high dominance, Brunner Mond were able to command more loyalty from their work force. Gone was the need for the harsh discipline of the earlier years, with its brutal fines and summary dismissals. Economic dependence ensured obedience, and from 1889 onwards Brunner's undoubted mastery of the art of management came to the fore. The open and crude oppressive exploitation of the early years gave way to more oblique forms of control, all the more effective for their subliminal nature.[124]

In 1880 Brunner Mond had 2,000 employees; by 1914 this had increased to 4,000, and by 1920 to 8,000; and since many small local firms were dependent upon them, perhaps three-quarters of the community of Northwich were dependent on them for their livelihood. Didsbury pointed out that Northwich has reflected the politics of the employer. Brunner was elected as a Liberal MP in 1885, and was succeeded by his son John Fowler Brunner in 1909. With the decline of the Liberals, Northwich became a safe Tory seat, and the local MP frequently reminded the electors in later years that the Labour party stood for the nationalization of ICI.[125]

A similar dependence was to be found by Roderick Martin and R. H. Fryer in their study of the effect of redundancies ordered by the major employer in 'Casterton' (Lancaster):

Casterton's geographical isolation, well-established commercial and kinship ties with surrounding rural areas, demographic stagnation, imbalanced occupational structure, low educational level, and emotional as well as economic dependence upon a limited number of employers, especially the Mills, sustained a complex structure of paternalistic capitalism, together with its supporting array of attitudes, until

[123] Lady Bell, *At the Works* (London, 1907), p. 23.

[124] Brian Didsbury, 'Cheshire Saltworkers' in R. Samuel (ed.), *Miners, Quarrymen and Saltworkers* (London, 1977), p. 183.

[125] Ibid., p. 184.

the redundancy. This was characterised by authoritarianism, tempered with generosity, on the part of the Mills, and deference, tinged with resentment, on the part of the employed.[126]

Like Northwich, and like the Dukeries mining villages, Lancaster voted in a much more conservative manner than its class and occupational structure might indicate, electing Tory MPs, often members of the employing family, in each general election in the twentieth century, except for 1966.

The 'new industry' towns of the twentieth century also demonstrate some of the typical patterns and problems of new communities. At Oxford, where the economy of the city was transformed by the arrival of the motor industry in the form of Morris Motors (Cowley) and Pressed Steel, working-class movements such as the Labour party and strong trade unions were founded for the first time. Many of those who worked in the motor industry were migrants from South Wales and other depressed areas, and indeed much of the impetus towards a new militancy amongst the workers in the 1930s stemmed from such immigrants, who took a lead in union activism, especially in the strike at Pressed Steel in 1934, when nine of the eleven members of the strike committee were long-distance migrants. However, Welsh immigration to the expanding new-industry towns did not always imply an influx of militancy. At Slough, where a less stable community was established, Welsh energies were diverted into leisure activites. There seems to have been little Welsh impact among Vauxhall car workers at Luton. Welshmen might respond to their new circumstances in other ways too; one third of the Welsh immigrants to Oxford soon returned to their homeland.[127]

In expanding towns where new industrial development took place between the wars, hostility between the existing residents and the newcomers was found. Margaret Stacey reported that in the case of Banbury there was a 'clash of

[126] R. Martin and R. H. Fryer, *Redundancy and Paternalist Capitalism* (London, 1973), p. 26.

[127] R. Whiting, 'The working class in the new industry towns between the Wars: the case of Oxford', Oxford D. Phil. 1978; J. Zeitlin, 'The Emergence of the Shop Steward Organisation and Job Control in the British Car Industry: a Review Essay', *History Workshop*, 10 (1980), 119–37.

cultures' between the immigrants and native Banburians.[128] At the glass and chemicals town of St. Helens in the twentieth century, the effects of the power of a single large employer can also be seen:

> Pilkingtons' dominance is dominance of the labour market. It is not just that no other employer of male labour can afford to pay very much less than Pilkingtons, it is also a question of the people's feeling of dependence upon Pilkingtons. Nothing can be done to offend the firm too much, for, who knows, they may pack their machines and take them somewhere else. In this situation people are likely to develop contradictory attitudes: an exaggerated deference on the one hand and a dull resentment on the other.[129]

Thus it may be concluded that many of the problems faced by the new Nottinghamshire coal-mining villages in the 1920s and 1930s were common to other new communities, whether they be twentieth-century housing estates and new towns or nineteenth-century factory villages and boom towns. There was a high rate of movement of population both in and out, a lack of adequate provision of services and amenities such as education, destruction of the spirit of kinship and community in favour of smaller 'family' units, and an emphasis on differing regional origins; and often a dominance on the part of the industry, the employer or the authority which brought the new community into being. This influence acted as a hindrance to the development of public participation in local affairs, and to popular control over the development of the town concerned. It was an influence often seen in wider aspects of life, such as in schools, churches, and local government, as well as 'at the works', and an influence reflecting the function of these communities as essentially an integral part of the manufacturing process rather than a blueprint for a balanced society, however 'benevolent' the paternalism involved. It is in this tradition, rather than as an idealistic attempt to secure 'a better future' that we should place the 'model villages' of the Dukeries coalfield.

[128] Margaret Stacey, *Tradition and Change, a Study of Banbury* (London, 1960), p. 14.

[129] A. Lane and K. Roberts, *Strike at Pilkingtons* (London, 1971), p. 31.

Conclusion

'Not as easy or as wise . . .'

Communities, like human beings, have teething troubles. Many of the distinctive features of the Dukeries mining communities of the 1920s and 1930s can be traced simply to their 'newness'. It is clear that colliery villages cannot be described in general terms without taking into account how long they have existed and whether they can be called mature and established communities. The high level of turnover of population in the Dukeries in the dozen or so years after the sinking of the pits represents a migration which in itself caused a considerable degree of disruption. This geographical mobility, not usually considered to be typical of coal-mining, presented problems in the planning and provision of facilities for education, leisure, sport, shopping, and female employment. In their early years the colliery villages demonstrated all the confused instability of the classic boom town, not only because they lacked the institutions and organizations of established townships but because of a more indefinable failure to enjoy what might be called 'community spirit'.

Miners came from the adjacent established coalfields of Nottinghamshire, Derbyshire, and South Yorkshire, and also, in smaller numbers, from distant declining fields such as the North East, Lancashire, and South Wales. There is evidence that regional customs and differences were seen to continue after the miners and their families had settled in the 'melting pot' of the Dukeries, and that it was no easy matter to create a unified spirit in the new villages. It was also not easy to bring the miners together with the existing inhabitants of the region, an area dominated by large landed estates and unused to rapid social and economic change. Although the aristocrats who still owned most of the land themselves reluctantly accepted the coming of mining to the Dukeries because of the financial attractions of coal royalties, there was an undercurrent of fear and dislike on the part of the 'old villagers' who resented the 'swamping' of their quiet culture by heavy industry and its workers. This resentment was still apparent after fifty years. The arrival of mining in the Dukeries may be seen as a case study in industrialization, in the effects of economic, social, and demographic change. It is particularly fortunate from the point of view of the richness of sources that this transformation took place recently enough to be within the memory of many participants, and to afford ample documentary evidence.

However, the youth of the mining villages does not explain all of their characteristics. Each community was planned by a single authority, the coal company engaged in exploiting the mineral resources of the locality. As a result they exhibited a level of social and economic homogeneity which classes them as occupational communities. They also suffered from an unusual concentration of power in the hands of the employers, whose political, economic, and social influence was all but overwhelming. This would, of course, have been impossible in a community which had developed piecemeal, with a variety of authorities, sources of employment, and owners of housing. Due to the nature of the Dukeries mining villages as 'company towns', certain unusual features may be observed. The representation of the villages in local government remained in the hands of the colliery management and the old leaders of Dukeries society, whereas the Labour party remained unorganized in the 1920s and 1930s. The role of trade union was fulfilled by the 'non-political' Spencer union, a Nottinghamshire phenomenon which benefited from almost a 'closed shop' in the new pits. The butty system remained in operation in several collieries. The miners' lives in the villages were beset by regulations. Sports and recreations, education, and religion were often provided by the grace of the management of the coal companies in their role as 'benevolent' paternalists. The companies adopted policies which were designed to attract labour to the pit and the colliery villages, but also to retain only 'good men'. They could simultaneously complain of a lack of reliable colliers and sack without hesitation any man who stepped out of line.

The concept of the 'model' village, with its neatness and

orderliness, emphasized and symbolized the order of the community, and its hierarchy was revealed by the gradation of standards of accommodation occupied by the inhabitants, ranging from the mansions of the resident colliery managers to the huts of the Irish sinkers. Every society does of course need a system of values which are passed on in common to its members in order to survive. But in the Dukeries colliery villages the mediation of this social control can be perceived in the painstaking policy of the employers. Determined to maintain these isolated villages as functional units separate from contaminating ideologies such as those purveyed by the Labour party and the Nottinghamshire Miners Association, the companies stressed the unity of the model villages. Outsiders such as door-to-door hawkers were regarded as threats, however minor, to the monopoly symbolized by the company store. The stability of the village community—represented by the presence of the colliery manager not only in the local manor house but at the head of ambulance parades and at sports events—was facilitated in the Dukeries by the small size of the social unit and the ability to maintain face-to-face contact between the agents of authority and the miners.[1] In larger communities it seems to prove harder to sustain such a relationship successfully.

It might be worthwhile to recall at this point that although the Nottinghamshire coalfield is often described as if it were a single homogeneous unit, moderate in politics and inclined to Spencerism and the butty system, many of the characteristics of the Dukeries field were not to be found in the older, more westerly mining communities. Around Kirkby and Sutton-in-Ashfield, for example, the influence of the employer was diluted by the proximity of a number of pits to each other, which offered the possibility of alternative employment to a miner threatened with dismissal. The proportion of company housing was very much lower, and it was quite impossible for a coal company to stand as paternalist provider of all goods and welfare facilities. The old western Nottinghamshire mining communities were not

[1] H. Newby, 'Paternalism and Capitalism: a Re-examination of the "Size-Effect"' in R. Scase (ed.), *Industrial Society; Class, Cleavage and Control* (London, 1977), pp. 59-73.

company towns. There was indeed a considerable variety of employment, for the area was well known for its textile manufacture, especially hosiery. In Sutton in 1921, only 51.5 per cent of occupied male workers were engaged in coal-mining. If women are included, the figure drops to 42.9 per cent. In Mansfield, which had a population in 1921 of 46,000, these figures are 44.6 and 37.3 per cent respectively.[2] It is scarcely surprising that it was quite possible to organize successful Labour parties in places like Sutton, and that there was always an effective NMA opposition to Spencerism in the western pits.[3] Yet as far as can be determined, there were few if any noticeable differences in the work process in the pit itself between old and new coalfields, apart from considerations of depth, working conditions, and modernity of equipment. Comparative work does therefore suggest that the newness and isolation of the Dukeries villages accounts for the unusual authority and command of the employers.

The colliery villages, built at the pit gates to enable the coal resources of the Dukeries to be tapped as efficiently as possible, can perhaps be seen as 'greedy institutions',[4] which sought undivided and exclusive loyalty from their members. Greedy institutions,

> though they may in some cases utilise the device of physical isolation, tend to rely on non-physical mechanisms to separate the insider from the outsider, and to erect symbolic boundaries between them. . . . Nor are greedy institutions marked by external coercion. On the contrary, they tend to rely on voluntary compliance and to evolve means of activating loyalty and commitment.[5]

The company town, the factory village and the pit village may all be seen as tending to extend the boundaries of the greedy institution to encompass a complete locality and to deny the members alternative definitions of their situation.[6] 'The factory and its neighbourhood reinforced the stability and totality of place, and made the identification of master and operative possible'.[7]

[2] 1921 Census, Nottinghamshire, Tables 16-17.
[3] See above, pp. 117-8.
[4] L. Coser, *Greedy Institutions* (New York, 1974).
[5] Coser, p. 6; Newby, in Scase (ed.), *Industrial Society*, p. 67.
[6] Newby, p. 68.
[7] Patrick Joyce, *Work, Society and Politics*, p. 94.

'Social order is maintained not only, or even mainly by legal systems, police forces and prisons, but is expressed through a wide range of social institutions from religion to family life.'[8] It may be true that in the case of the Dukeries that despite the extensions of company influence from the pit into the village, the wage relationship itself is the greatest social control. The monopoly of employment was the companies' ultimate strength, the threat of dismissal their ultimate weapon. Stedman Jones may well point out that the exercise of social control is merely the expected safeguarding of interests by powerful people or a hegemonic system, so the concept is open to the accusation that it is 'circular or meaningless'.[9] Nevertheless it may have use as a tool of clarification rather than an explanatory theory.[10]

It is fair to point out that the new Dukeries communities did not lose their distinctive nature until the 1940s, when the economic power of the employers was weakened by war regulations and controls, followed by nationalization of the coal industry, when the Spencer union gave way to the Nottinghamshire Area of the NUM, when the Labour party was founded, and swept to victory in the colliery villages, and when the aristocrats departed from the Dukeries, their estates atomized. It was not just a matter of waiting for time, of waiting for the villages to mature into the typical pattern of colliery communities.

The first lesson to be learnt from the history of the new Dukeries coalfield is that it is neither as easy nor as wise to attempt to plan new communities from scratch, and from above, as may be thought. The wide ranging and painstaking care of the highly efficient coal companies in protecting their substantial investment in the deep and profitable new undertakings may in a sense be admired. But it is not clear that the colliery villages were happy or fulfilling places in which to live. The way that miners and their families 'voted with their feet' in leaving the new villages in great numbers, despite the

[8] A. P. Donajgrodzki, *Social Control in Nineteenth Century Britain* (London, 1977), p. 9.

[9] G. Stedman Jones, 'Class expansion versus social control', *History Workshop*, 4 (1977), 164.

[10] H. F. Moorhouse, 'A Reply to Reid', *Social History*, 4 (1979), 484.

economic attractions of a secure and relatively well-paid job, may be added to the testimony of those who remained.

The Dukeries colliery villages of the 1930s may be criticized from a radical point of view which would stress the powerlessness of the miners, their inability to organize in effective political or industrial institutions, and that working conditions and pay were not adequate reflections of the contribution of the miners to the tapping of the spoils that the Dukeries had to offer. But the way the colliery villages were run may also be criticized from a liberal, pluralist point of view. The 'industrial feudalism' of the company villages, the lack of separation between work and non-work, was not to be the blueprint for the development of British society in the twentieth century. In this sense the model communities looked back to the factory villages of the Victorian paternalist entrepreneurs. It was from outside the Dukeries that the control of the private companies was destroyed, by changing circumstances and changing ideas about how Britain should be governed. Butterley and Bolsover, Barber-Walker and Stanton, failed in their attempts to stave off the threats to their lucrative exploitation of the minerals which lay beneath Sherwood Forest—but they were defeated by outside agencies.

Social, economic, and demographic change is constantly taking place. More specifically, new coalfields are regularly opened up. It is perhaps valid to base on the experience of the Dukeries coalfield a plea to aim where possible for mixed, balanced communities, to build on the established facilities and community life of existing towns and villages enjoying a multiplicity of authorities. By equalizing the distribution of power, privilege, participation, and opportunity, true democracy may perhaps be made possible. Thus some of the hardships, errors, and problems which beset Ollerton, Edwinstowe, Bilsthorpe, Harworth, and the other colliery villages may be avoided.

Note on Oral Evidence

A major source for the history of the new Dukeries coalfield lies in the oral evidence of the men and women who experienced the transformation: inhabitants of the old agricultural villages who can remember the district before the arrival of mining, as well as the pioneer colliers and their families. Needless to say, one must apply to oral evidence the same critical standards as to any other historical source. In particular, testimony about specific dates and events should be tested where possible against written records, for although local people (especially miners) often have a keen interest in their own history and traditions, and a good memory of the major developments in their communities, the passing of the years can lead to inaccuracy and distortion. This is one of the reasons why I prefer to talk of 'oral evidence', one among many types of sources, rather than 'oral history'.

Indeed, it may well be that people cannot 'write their own history' with the degree of analysis and insight that one might wish, for their perceptions of society, whether local or national, are influenced by the role they play in it and by the power distribution of that society itself. Views and values can, perhaps must, be inculcated in order to encourage and safeguard stability and quiescence. This interpretation of the history of the young Dukeries mining communities must be my own. It involves an implication that those who accepted in full the concept of the 'model' village were victims of a consensual view, spread by the employers but not necessarily originated by them. This view happened to coincide with the management's interests in preserving social and industrial harmony and continuity of production. Even those interviewees who resented the power of the companies often did not question more fundamental social and political shibboleths. On the other hand, the historian is bound to introduce his own biases and values, as the use of any kind of evidence necessarily involves selection and editing on his part, which creates the order of interpretation.

Finding interviewees also poses problems. Although in theory the complete set of the inhabitants of the villages who lived through the social and economic changes of the inter-war years can be traced by a comparison of electoral registers through the years, even so this procedure could not enable us to build up a representative sample for interview. What of the many residents of the villages who left the Dukeries years ago and are now virtually untraceable? In practice, the method adopted was that of establishing initial contacts through a variety of channels—advertising in local newspapers, and approaching business, trade union, and family contacts, and then expanding the field of inquiry as each contact introduced friends and acquaintances. As long as care is taken not to become confined within any clique or interest-group, this can prove a satisfactory technique which can provide a range of views and experiences of the industrialization of the Dukeries: old village and new, management and worker, wives and children.

Despite the problems concerning the procurement and reliability of interviewees, it would be foolish to ignore the treasury of experience of the participants in the social and economic revolution in the Dukeries. It is true that no statistical weight can be placed on the answers obtained to questions as it could in a contemporary social survey which can apply a scientific sampling technique. But then a successful interview should not follow the lines of a rigid questionnaire. After all, the greatest value of interviews is not to elicit specific factual information but to acquire a better judgement of the attitudes and ambience of a period, to test the historian's assumptions, and, when quoted, to add vividness and colour to the text.

For example, the hostility between 'old' and 'new' villagers hinted at in the local newspapers is illustrated and illuminated by several interviews. The reasons for the lack of Labour party activity in the mining villages before the Second World War can only be guessed at from the municipal election results which showed that Conservative and Independent candidates were returned unopposed in most years. Oral evidence gives some clues as to the explanation: the fear of reprisals against those who dared to oppose in public the management nominees. Often oral testimony can provoke a fresh line of inquiry, and enable the historian actually to indulge in an ambition which has become a cliché, the 'dialogue with his sources'. The reactions to the questions concerning the impact of nationalization on the Dukeries colliery villages provided an at first unexpected insight: 'freedom' from company control might be counterbalanced by the notion that the rot had set in as far as the order, discipline, and 'pride' of these almost unique model communities were concerned. In an interview, a balance must be sought between allowing the interviewee to ramble and introducing too much interference on the part of the interviewer. Similarly, the historian's interpretation must be shaped by what he hears, although clearly his own interests and ideas will influence his choice of questions.

Finally, oral evidence can provide descriptions of aspects of life in a colliery village which are relatively inaccessible through documentary material: time spent at leisure rather than at work, entertainment, shopping, social *mores*. The testimony of those who passed through the education system at the time can be gleaned, although we must remember that for the 1920s at least, oral evidence tends to a 'child's eye view' of the world. It is important to consult also the schools admissions registers and log books deposited in the Nottinghamshire County Record Office and the Schools Inspectors Reports in the Public Record Office at Kew.

The Dukeries coalfield in the inter-war years is a subject circumscribed enough in time and space to permit an attempt at the reconstruction of the 'total' history of the communities and their population, not just of their leaders, and their leaders' activities and decisions, as history based on 'official' sources only might stress. Oral evidence is an essential tool in the pursuit of such an ambition.

Bibliography

1. Primary sources

In Derbyshire County Records Office, Matlock

NCB N1-N5 Butterley Company records.
NCB N6-N8 B1-21 Bolsover Colliery Company records.
B22-43 Blackwell Colliery Company records.

At British Steel, Stanton

Stanton Company Management Committee Minute Books
Collieries Committee Minute Books
Minutes of the Board
A. E. G. Harmon, History of Stanton (8 vols., 1960, unpublished typescript).

At the collieries

Signing-on books: Harworth Colliery 1924-31
Bilsthorpe Colliery 1926-41
Ollerton Colliery 1923-51
Blidworth Colliery 1941-8
Clipstone Colliery 1915-23

In the possession of Dr A. R. Griffin

Thoresby Colliery Manager's Annual Reports 1925-51.
Nottinghamshire Miners Industrial Union (Spencer union), Rufford Branch Minutes 1926-37.

In Nottinghamshire County Records Office, Nottingham

SA Schools Admissions and Leavers Registers:
Harworth Bircotes Junior Infants Mixed
Harworth Bircotes North Border Council School
Ollerton Wellow Road
Blidworth Wesleyan
Hodsock Langold Temporary
Kirkby-in-Ashfield East Council
Mansfield Woodhouse Oxclose Lane Council
Sutton-in-Ashfield Central Infants
Hucknall Upper Standard Mixed.
ED CC Nottinghamshire Education Committee Notes of Visit.

CC County Council records:

CL Clerk's Department Registration of Cinemas.
Highways Committee Minute Book 1915-31.

DC/SW Southwell Rural District Council records:

Southwell RDC rate books.
Southwell RDC North Highways Committee Minute Books.
Southwell RDC signing-in book for Councillors.
Register of buildings for solemnizing marriages.

PR Parish records:

Edwinstowe Parish Register of Baptisms
Ollerton Parish Banns of Marriage.

BP Ollerton Baptist Church Minute Book 1926-60.

PUS Southwell Union Poor Law Minutes.

PP Broxtowe Constituency Labour Party Council Minutes 1925-44; Executive Committee Minutes 1917-44.

Electoral registers: Bassetlaw, Broxtowe, Mansfield, Newark constituencies.

In the Public Records Office, Kew

ED 21 School Inspector's Reports:

Edwinstowe Voluntary
Bilsthorpe Council
Harworth Bircotes County
Edwinstowe Council
Ollerton Whinney Lane
Blidworth Council
Clipstone Council
Blidworth C. of E.
Hodsock Langold Council
Hucknall Beardall Street Council
Heanor Loscoe Road Council
Eastwood Council.

Nottinghamshire Premises Survey 1925.

COAL 12 Reorganization Committee.

LAB 23 Report on Transferees from Special Areas.

HLG 4 Southwell RDC Town Planning Scheme 1923.

In the possession of Mr Miles Wright

M. F. M. Wright, Memoirs (unpublished).

In Nottingham University Library, Manuscripts Department

Manvers of Thoresby Papers
Portland of Welbeck Papers

Middleton Papers
Galway of Serlby Papers.

In Nottinghamshire Area NUM Library, Berry Hill Lane, Mansfield

Nottinghamshire Miners Association Minute Books 1919-36.
Nottinghamshire Miners Federated Union:
Minute Books 1939-45;
Political Minute Books 1937-47.
National Union of Mineworkers, Nottinghamshire Area:
Minute Book 1948;
Branch Returns 1947-51;
Political Minutes 1947-51.

In Bassetlaw Labour Party Headquarters, Worksop

Bassetlaw Constituency Labour Party Minute Books 1932-5.

2. Newspapers

Bircotes Magazine (Harworth) 1924-5
Colliery Guardian (London)
Doncaster Gazette
Dukeries Advertiser (Ollerton) 1934-8
Mansfield and North Notts Advertiser (Mansfield)
Mansfield Chronicle
Mansfield Reporter and Sutton Times (Mansfield)
Ministry of Labour Gazette (London)
Newark Advertiser
Nottingham Evening Post
Nottingham Guardian
Nottingham Journal
Ollerton, Edwinstowe and Bilsthorpe Echo (Worksop) 1941-5
Ollerton, Edwinstowe and Bilsthorpe Times (Ollerton) 1938-41
Retford, Gainsborough and Worksop Times (Retford)
Worksop Guardian
Kentish Observer (Canterbury).

3. Interviews

Nearly 100 people were interviewed, but only extracts from the following are included in the text.

Baker, Ernest H., 21 Aug. 1979, Langold. Born Cambridge, 1920, son of a haulage hand (Shireoaks Colliery, Worksop) turned military prison medical warder (1914-22). Returned to Shireoaks 1922, father assisted with sinking of Firbeck Main Colliery from 1924. Moved into colliery village of Langold 1925. Ed. Langold Council and Worksop Central School, left 1935 to become a pony driver underground, then office boy. Telephone operator during war, then returned to Firbeck. Rose to become Chief Clerk (Admin. Officer) of Firbeck, and on its closure in 1968 Admin. Officer of Shireoaks Colliery (to date). COSA and APEX union branch secretary, Parish Council, active member of the Labour party and Langold C. of E. Church.

Banks, John Edward, 5 Aug. 1979, Bircotes. Born Askern, Yorkshire, 1924. Came to Harworth 1936. Ed. Bircotes North Border Council. Mrs Banks lived in Bircotes all her life, her parents coming from Rossington and Thorne, both in Yorkshire.

Booth, Harold, 21 Aug. 1979, Langold. Born 1910. When he came to work at Firbeck (1928), he had already worked at seven pits in Nottinghamshire and Yorkshire–first Rossington (1924). Nephew of Sam Booth, NMA activist and checkweighman at Ollerton, ousted in 1930 in favour of the Spencer union's Aaron Jenkins. Deputy at Firbeck. NUM branch chairman. Member Worksop RDC (Labour–professed Bevanite). Secretary or Treasurer of Langold Labour party for thirty-one years. Former JP. Retired.

Buxton, Barbara, 3 Jan. 1979, Forest Town, Mansfield. Moved to Edwinstowe 1930 when still a baby, remained there until 1972. Ed. King Edwin's Council School.

Buxton, Cyril, 3 Jan. 1979, Forest Town, Mansfield. Born 1920, Warsop, Notts. Lived in Ollerton 1926-42. Ed. Whinney Lane Council School, Ollerton, Brunt's GS, Mansfield, Miners Welfare Exhibition to King's College, London University 1938. Schoolteacher. Son of a butty.

Cocker, George, 22 Dec. 1978, Edwinstowe. Worked for many years for Percy Morley, garage proprietor.

Corke, Alfred Edward, 3 Aug. 1978, Walesby near Ollerton. Born 1908, came to Ollerton 1927 to operate projector in picture house from Matlock, Derbyshire. Hall and Earl's textile factory 1937. Pit electrician 1969. Retired 1975. Southwell RDC Councillor for Ollerton (Labour) in early 1950s.

Davies, Carrie, 6 Aug. 1979, Edwinstowe. Born 1912. Came from Creswell, Derbyshire to Ollerton, 1926. To Edwinstowe 1927. Worked at Woolworth's in Hyson Green, Nottingham for seven years, cycling to and from work.

Green, Charles Henry, 3 Aug. 1978, New Ollerton. Lived in Ollerton since 1926. Father worked at colliery from sinking. Worked as apprentice to butcher in Retford, then at Ollerton Brickworks, and from 1927 at pit. Retired 1974.

Guest, Joseph Hayward, 6 Aug. 1978, Clipstone. Born 1902. Came from Pilsley, Derbyshire to Ollerton Colliery in 1927, and moved to Thoresby Colliery in 1931. Worked at coal-face until 1937, deputy 1937-55, overman 1955 until retirement in 1965. Formed a branch of NACODS in 1946 and was branch secretary for sixteen years.

Hawkins, Neville, 30 June 1979, Bircotes. Born 1915. Originally from Grassmoor Colliery, NE Derbyshire. Worked at Harworth Colliery from 1931, living in Tickhill. Moved into Bircotes Colliery village in 1937. Became Labour Councillor for Harworth (Worksop RDC) in 1949. Retired.

Howard, Mrs, 2 Aug. 1978. Now lives in Johannesburg, South Africa. Born 1909. Father a miner at Welbeck Colliery, Nottinghamshire. Lived in Mansfield until 1929, when she emigrated to South Africa to join two brothers. Aunt of Miss Hilda Tagg of Rainworth (see below).

Jackson, Jim, 21 Aug. 1979, Langold. Official at Firbeck Main. Retired. Lived at Carlton in Lindrick in the 1920s.

Jenkins, Thomas Henry, 30 July 1979, Harworth. Born 1908, Penistone. Father a sinker at Rossington, then at Harworth. Worked at pit from age fourteen. Retired.

Marshall, Frederick Sherman, 6 Aug. 1979, New Ollerton. Born Lincolnshire Wolds, came to Ollerton 1925 to work as handyman at Plough public house in New Ollerton. Drove coaches for colliery sports teams on Saturdays for many years. Special Constable during war.

Morley, Mabel, 4 Jan. 1979, Bilsthorpe. Born Mansfield Woodhouse, moved to Boughton near Ollerton at the age of three. Married a miner at Ollerton Colliery. Moved to Bilsthorpe in 1941. Widow.

Morris, Stanley, 30 July 1979, Harworth. Born 1919, Beighton on Derbyshire-Yorkshire border. Father came to work at Harworth Colliery 1924. Worked at pit since age fourteen, now on maintenance staff.

Nuttall, Albert, 3 Aug. 1979, Ollerton. Born 1894. Moved to Ollerton from Budby village on Thoresby Estate in 1924. Worked for Colemans of East Kirkby, contractors who built Ollerton colliery village. Retired.

Parnell, Harry, 21 Dec. 1978, Edwinstowe. Born 1892. Lived in the old village all his life. Mrs Parnell originally from Doncaster.

Slack, Samuel, 21 Aug. 1979, Langold. Lived in Langold all his life. Worked at Firbeck then Manton Collieries. Sitting councillor for Langold on Bassetlaw District Council (Labour). Former JP. Retired.

Smith, Gersh, 3 Aug. 1979, New Ollerton. Born 1912, Sutton-in-Ashfield, Nottinghamshire. Came to Ollerton Colliery Nov. 1928. Parish councillor (Labour), Boughton, from 1946. Retired 1977.

Smith, Joseph Edwin, 3 Aug. 1978, New Ollerton. Born Forest Town, Mansfield. Worked at Crown Farm Colliery, Mansfield. Came to Ollerton 1927 as a miner. Became Clerk to Ollerton Parish Council.

Spencer, John H., 3 Aug. 1979, New Ollerton. Born 1909, South Normanton, Derbyshire. Father came to work at Ollerton Colliery June 1927. Weigh clerk at Ollerton all his working life. In RASC during war. Nephew of George Spencer, founder of the NMIU. Retired. Widower.

Sperry, Wilfred, 6 Aug. 1979, Ollerton. Born Warsop Vale. Worked as miner at Warsop Main, Gedling, Bilsthorpe, and Ollerton Collieries. Early Labour party member in Ollerton.

Stringer, Charles Edward, 20 July 1979, Bircotes. Came from London in 1924 to help first curate, Revd Percy Leeds, with youth work. Labour councillor after the war. Retired.

Tagg, Hilda, 2 Aug. 1978, Rainworth. Lived at Rainworth since May 1932. Father worked at Rufford Colliery from 1923, living at Blidworth 1923-32. Previously he worked at Welbeck Colliery.

Tuck, Herbert Edward Charles, 4 Jan. 1979, Bilsthorpe. Came to Bilsthorpe 1927. Father born in Somerset, where he started work at the age of twelve, and subsequently worked in South Wales coal mines and at Stanton Hill, Sutton in Ashfield before coming to Bilsthorpe. Ed. Bilsthorpe Council School. Worked at Bilsthorpe Colliery from age fourteen. Elected 1946 as one of the first Labour parish councillors for Bilsthorpe, and sitting councillor on Newark District Council.

Tyndall, Charles, 21 Dec. 1978, Edwinstowe. Lived at Edwinstowe from age of two. Worked on railways. Brother of F. W. Tyndall.

Tyndall, Dorothy, 22 Dec. 1978, Edwinstowe. Wife of Frank Tyndall. Her family came from Tibshelf. Father moved from Blackwell Colliery to Thoresby at the beginning of 1931.

Tyndall, Frank William, 22 Dec. 1978, Edwinstowe. Lived at Edwinstowe from age of four. Father operated post office in old village of Edwinstowe. Worked at pit.

Wilson, Marjorie, 6 Aug. 1978, Mansfield. Father worked at Blidworth Colliery, lived in Blidworth colliery village and later at Ravenshead. Previously he was a sinker at Clipstone Colliery. Marjorie works for NCB as Secretary at Clipstone Colliery. Single.

Wright, Fr. Stephen, 13 Aug. 1982, Ampleforth Abbey. Eldest son of Ollerton Colliery Manager Monty Wright. Born Ollerton Hall 1937. Ed. Ampleforth and St. Benet's Hall, Oxford. OSB, Ampleforth Abbey.

4. Theses

Burkett, A. 'The committee system in selected local authorities in the East Midlands'. Nottingham MA 1960.

Day, M. G. 'Sir Raymond Unwin (1863-1940) and R. Barry Parker (1867-1947): a study and evaluation of their contribution to the development of site-planning theory and practice'. Manchester MA 1973.

Griffin, A. R. 'The development of industrial relations in the Nottinghamshire coalfield'. Nottingham Ph.D. 1963.

Hornsby, K. A. 'The geographical aspects of changing economic patterns with special reference to East Nottinghamshire and West Cumberland'. Nottingham Ph.D. 1958.

Jennings, J. R. G. 'Mansfield, the evolution of a changing landscape'. Nottingham Ph.D. 1966.
Johnson, W. 'The development of the Kent coalfield'. Kent Ph.D. 1972.
Law, C. M. 'Aspects of the economic and social geography of the Mansfield area'. Nottingham M.Sc. 1961.
Long, B. F. L. 'A study of the membership of selected local authorities with specific reference to social and political change'. Nottingham MA 1964.
Mason, L. M. E. 'Industrial development and the structure of rural communities'. Nottingham M.Sc. 1966.
Morris, G. M. 'Primitive Methodism in Nottinghamshire 1815-1932'. Nottingham Ph.D. 1967.
Shorter, P. R. 'Electoral politics and political change in the East Midlands of England 1918-35'. Cambridge Ph.D. 1975.
Smith, D. M. 'The East Midlands area: a regional study of industrial location'. Nottingham Ph.D. 1961.
Summers, N. 'Problems of visual environment in Nottinghamshire'. Nottingham Ph.D. 1966.
Taylor, R. C. 'The implications of migration from the Durham coalfield: an anthropological study'. Durham Ph.D. 1966.
Whiting, R. C. 'The working class in the new industry towns between the Wars: the case of Oxford'. Oxford D.Phil. 1978.

5. Secondary

(a) Nottinghamshire

Banks, J. A. *An Outline History of St. Giles and St. Paulinus Churches, Ollerton* (Ollerton, n.d.)
Bramley, J. 'Harworth', *Nottinghamshire Countryside*, April 1939.
Brown, W. H. *Mansfield's Cooperative Advance 1864-1950* (Mansfield, 1950).
Craig, F. W. S. *British Parliamentary Election Results 1918-49* (Chichester, 1969).
— *British Parliamentary Election Results 1950-70* (Chichester, 1971).
— *Boundaries of British Parliamentary Constituencies* (Chichester, 1972).
Lewis, C. Day. *The Buried Day* (London, 1960).
Edwards, K. C. *Nottingham and its Region* (Nottingham, 1966).
Ellis, W. L. 'The Miners' Struggle in Notts', *Labour Monthly*, 18 (1936), 34-9.
Graves, C. 'A Coal Miner's Life Above Ground', *The Sphere*, 23 April 1932, pp. 140-1, 162.
Gregory, R. *The Miners and British Politics* (Oxford, 1968).
Griffin, A. R. *The Miners of Nottinghamshire 1914-44* (London, 1962).
— *Mining in the East Midlands 1550-1947* (London, 1971).

Griffin, A. R. and C. P. Griffin. 'The Non-Political Trade Union Movement' in A. Briggs and J. Saville (eds.), *Essays in Labour History vol. iii 1914–39* (London, 1977) pp. 133–62.

Kennedy, P. A. *Guide to the Nottinghamshire County Records Office* (Nottingham, 1960).

Kidd, R. *The Harworth Colliery Strike*, a report to the Executive Committee of the National Council of Civil Liberties by the Secretary of the Council (1937).

Leeman, F. W. *Cooperation in Nottingham* (Nottingham, 1963).

Mines Department. *Annual List of Mines* 1921–38, 1945–51.

Mottram, R. H. and C. Coote. *Through Five Generations: the History of the Butterley Company* (London, 1950).

Municipal Yearbook 1920–51.

Murphy, C. *Harworth Colliery 1924–74* (Harworth, typescript, 1974).

New Statesman. 'The New Coalfield in Nottinghamshire', 24 Dec. 1927.

Nottinghamshire Community Project Foundation. *Cotgrave Report* 1978.

Portland, Duke of. *Men, Women and Things* (London, 1937).

Powell, A. G. 'The 1951 census: an analysis of population changes in Nottinghamshire', *East Midland Geographer*, 2 (1954), 13–22.

Social Services Department, Rushcliffe District Council. *Cotgrave* (1977).

Taylor, Lord. *Uphill all the Way* (London, 1972).

Tomkinson, D. 'East Midland Mining Communities', *Journal of the Town Planning Institute*, 43 (1957), 85–8.

Townroe, B. S. 'A Miner's Village', *The Spectator*, 10 April 1926.

Walters, Sir J. Tudor. *The Building of Twelve Thousand Houses* (London, 1927).

Whitelock, G. C. H. *250 Years in Coal, the History of the Barber-Walker Company 1680–1946* (Derby, 1955).

Who's Who in Nottinghamshire (Worcester 1935).

(*b*) *Other coalfields*

Abercrombie, P. and J. Archibald. *East Kent Regional Planning Scheme* (2 vols.) (Liverpool and Canterbury, 1925, 1928).

— and T. H. Johnson. *Doncaster Regional Planning Scheme* (Liverpool, 1922).

Atkinson, F. *The Great Northern Coalfield 1700–1900* (Barnard Castle 1965).

Bean, D. 'Kent and its Miners', *Coal Quarterly*, 1 (1963), 8–10.

Berkovitch, I. *Coal on the Switchback* (London, 1977).

Birch, T. W. 'Development and decline of the Coalbrookedale coalfield', *Geography*, 19 (1934), 114–26.

Bulmer, M. I. A., (ed.), *Mining and Social Change: Durham County in the 20th Century* (London, 1978).

— 'Sociological Models of the Mining Community', *Sociological Review*, NS 23 (1975), 61–92.

Buxton, N. K. 'Entrepreneurial efficiency in the British Coal Industry between the Wars', *Economic History Review*, 2 s 23 (1970), 476–97.
Campbell, A. *The Lanarkshire Miners* (Edinburgh, 1979).
Challinor, R. *The Lancashire and Cheshire Miners* (Newcastle, 1972).
Colliery Guardian. *Guide to the Coalfields 1979* (Redhill).
Court, W. H. A. *Coal* (London, 1951).
Davies, W. H. *The Right Place, the Right Time* (Llandybie, 1972).
Davison, J. *Northumberland Miners 1919–39* (Newcastle, 1973).
Dennis, N., E. Henriques, and C. Slaughter. *Coal is Our Life* (London, 1956).
Down, C. G. and A. J. Warrington. *The History of the Somerset Coalfield* (Newton Abbot, 1971).
Garside, W. R. *The Durham Miners 1919-60* (London, 1971).
Gaventa, J. P. *Power and Powerlessness: Quiescence and Rebellion in an Appalachian Valley* (Oxford, 1980).
Gray, G. D. B. 'The South Yorkshire Coalfield', *Geography*, 32 (1947), 113–31.
George, D. Lloyd. *Coal and Power* (London, 1924).
Goffee, R. E. 'The Butty System and the Kent Coalfield', *Bulletin of the Society for the Study of Labour History*, 34 (1977), 41-55.
Griffin, A. R. *The British Coalmining Industry* (Hartington, 1977).
Griffiths, I. L. 'Zambia's new coalfields', *Geography*, 53 (1968), 415–18.
Harkell, G. 'The Migration of Mining Families to the Kent Coalfields between the Wars', *Oral History*, 6 (1978), 98-113.
Hart, C. *The Industrial History of Dean* (Newton Abbot, 1971).
Hassan, J. A. 'The Landed Estate, Paternalism and the Coal Industry in Midlothian, 1800-1880', *Scottish Historical Review*, 59 (1980), 73-91.
Haynes, W. W. *Nationalization in Practice: the British Coal Industry* (London, 1953).
House, J. W. and E. M. Knight. 'Pit Closure and the Community', *Papers in Migration and Mobility in Northern England*, University of Newcastle upon Tyne, 5 (1967).
Jevons, H. S. *The British Coal Trade* (1915, reprinted Newton Abbot 1969 with an introduction by B. F. Duckham).
Johnson, W. 'A second comment', *Economic History Review*, 2 s 25 (1972), 665–8.
Jones, P. N. 'Colliery Settlement in the S. Wales Coalfield', University of Hull *Occasional Papers in Geography*, 14 (1969).
Kirby, M. W. 'Comment', *Economic History Review*, 2 s 25 (1972), 655–657.
— 'The control of competition in the British coal mining industry in the Thirties', *Economic History Review*, 2 s 26 (1973), 273–84.
— *The British Coalmining Industry* (London, 1977).
Knight, E. M. 'Men leaving Mining, West Cumberland 1966–7', University of Newcastle upon Tyne, *Papers on Migration and Mobility in Northern England*, 6 (1968).
Lawson, J. *A Man's Life* (London, 1932; 1944).

Lebon, J. H. G. 'The development of the Ayrshire coalfield', *Scottish Geographical Magazine*, 49 (1933), 138-53.

Loubère, L. 'Coal Miners, Strikes and Politics in the Lower Languedoc, 1870-1914', *Journal of Social History*, 2 (1968-9), 25-50.

Macfarlane, J. 'Denaby Main: a South Yorkshire Mining Village', *Bulletin of the Society for the Study of Labour History*, 25 (1972), 82-100.

Moore, R. S. *Pitmen, Preachers and Politics* (Cambridge, 1974).

Moyes, A. 'Post-war Changes in Coalmining in the West Midlands', *Geography*, 59 (1974), 111-20.

Murphy, R. E. 'A Southern West Virginia Mining Community', *Economic Geography*, 9 (1933), 51-9.

Neuman, A. M. *The Economic Organization of the British Coal Industry* (London, 1934).

North, J. and D. Spooner. 'On the coal mining frontier', *Town and Country Planning*, 46 (1978), 155-63.

Neville, R. G. and J. Benson. 'Labour in the coalfields (II), a select critical bibliography', *Bulletin of the Society for the Study of Labour History*, 31 (1975), 45-59.

Peace, K. 'Some changes in the S. Yorkshire coalmining industry 1951-71', *Geography*, 58 (1973), 340-2.

Raybould, T. J. *The Economic Emergence of the Black Country* (Newton Abbot, 1971).

Redmill, C. E. 'The growth of population in the E. Warwickshire coalfield', *Geography*, 16 (1931), 125-39.

Riley, R. C. 'Recent developments in the Belgian Borinage: an area of declining coal production in the European Coal and Steel Community', *Geography*, 50 (1965), 261-73.

Smailes, A. E. 'Population changes in the colliery districts of Northumberland and Durham', *Geographical Journal*, 91 (1938), 220-32.

Smith, D. 'The struggle against Company Unionism in the S. Wales coalfield 1926-39', *Welsh History Review*, 6 (1973), 354-78.

Storm-Clark, C. 'The Miners 1870-1970, a test case for Oral History', *Victorian Studies*, 15 (1971-2), 49-74.

Thomas, B. 'Labour mobility in the S. Wales and Monmouthshire coalmining industry 1920-30', *Economic Journal*, 41 (1931), 216-26.

Thompson, E. P. Review of R. S. Moore, *Pitmen, Preachers and Politics*, *British Journal of Sociology*, 27 (1976), 387-402.

Thompson, I. B. 'A geographical appraisal of recent trends in the coal basin of Northern France', *Geography*, 50 (1965), 252-60.

Turner, P. A. 'Colliery development in the inter-war period, the opening of the Markham Collieries, Derbyshire 1924-30', *Derbyshire Archaeological Journal*, 98 (1978), 86-9.

Watkins, H. M. *Coal and Men* (London, 1934).

White, P. H. 'Some aspects of urban development by colliery companies 1919-1939', *Manchester School of Economics and Social Studies*, 23 (1955), 269-80.

Williams, J. E. *The Derbyshire Miners* (London, 1962).

Wise, M. J. 'Some notes on the growth of population in the Cannock Chase coalfield', *Geography*, 36 (1951), 235–48.
Zweig, F. *Men in the Pits* (London, 1948).

(*c*) *Parallels and theory*

Alonso, W. 'What do we have new towns for?', *Urban Studies*, 7 (1970), 37–55.
Ashworth, W. *The Genesis of Modern British Town Planning* (London, 1954).
Bailey, P. *Leisure and Class in Victorian England* (London, 1978).
Bell, C. and H. Newby. *Doing Sociological Research* (London, 1977).
Birch, A. H. *Small Town Politics: a Study of Political Life in Glossop* (London, 1959).
Bournville Village Trust. *When We Build Again* (London, 1941).
Brennan, T. *Reshaping a City* (Glasgow, 1959).
Briggs, A. *Victorian Cities* (London, 1963).
Buder, S. *Pullman, an Experiment in Industrial Order and Community Planning 1880–1930* (New York, 1967).
Bulmer, M. I. A. (ed.), *Working Class Images of Society* (London, 1975).
Burke, G. *Towns in the Making* (London, 1971).
Caillois, R. *Man, Play and Games* (Paris, 1958; London, 1962).
Chaloner, W. H. *The Social and Economic Development of Crewe 1780–1923* (Manchester, 1950).
Clayre, A. *Work and Play* (London, 1974).
Collison, P. *The Cutteslowe Walls* (Oxford, 1963).
Crossick, G. *An Artisan Elite in Kentish London 1840–1880* (London, 1978).
Daly, M. 'Reply', *Sociological Review*, 30 (1938), 236–61.
Daniel, G. H. 'Labour Migration and Age Composition', *Sociological Review*, 30 (1938), 281–308.
— 'Labour Migration and Fertility', *Sociological Review*, 31 (1939), 370–400.
— 'Some Factors affecting the Movement of Labour', *Oxford Economic Papers*, 3 (1940), 144–79.
Day, M. G. and K. Garstang, 'Socialist theories and Sir Raymond Unwin', *Town and Country Planning*, 43 (1975), 346–9.
Dewhirst, R. K. 'Saltaire', *Town Planning Review*, 31 (1960–1), 135–44.
Dore, R. P. *British Factory Japanese Factory* (London, 1973).
Durant (Glass), R. *Watling* (London, 1939).
Epstein, A. L. *Politics in an Urban African Community* (Manchester, 1958).
Goldthorpe, J. H., *et al. The Affluent Worker in the Class Structure* (London, 1968–9).
Gray, R. Q. *The Labour Aristocracy in Victorian Edinburgh* (Oxford, 1976).

Hareven, T. K. and R. Langenbach. *Amoskeag* (London, 1979).
Harrison, M. 'Burnage Garden Village: an ideal for life in Manchester', *Town Planning Review*, 47 (1976), 256-68.
Havinden, M. A. *Estate Villages* (London, 1966).
Heraud, B. J. 'Social Class and the New Towns', *Urban Studies*, 5 (1968), 33-58.
Holroyd, A. *Saltaire and its Founder* (London, 1871).
House, J. W. 'The Applied Geographer and the Steady State Society: the case of the New Towns', *GeoJournal*, 1 (1977), 15-24.
Howard, E. *Garden Cities of Tomorrow* (London, 1898, 1902, 1946 with preface by F. J. Osborn and introductory essay by L. Mumford).
Hudson, R., M. R. D. Johnson, *et al. New Towns in North East England* (2 vols., Durham, 1976).
Jackson, J. A., (ed.), *Migration* (Cambridge, 1969).
Jennings, H. *Societies in the Making* (London, 1962).
Jessop, R. D. 'Culture and Traditionalism in English Political Culture', *British Journal of Political Science*, 1 (1971), 1-24.
Jevons, R. and J. Madge, *Housing Estates* (Bristol, 1946).
Jewkes, J. and H. Campion. 'The Mobility of Labour in the Cotton Industry', *Economic Journal*, 38 (1928), 135-7.
Joyce, P. *Work, Society and Politics* (Brighton, 1980).
Kuper, L. (ed.). *Living in Towns* (Birmingham, 1953).
Lane, A. and K. Roberts. *Strike at Pilkingtons* (London, 1971).
Lockwood, D. 'Sources of variation in working class images of society', *Sociological Review*, NS 14 (1966), 249-67.
Lukes, S. *Power* (London, 1974).
Makower, H., J. Marschak, and H. W. Robinson. 'Studies in the Mobility of Labour, a tentative statistical measure', *Oxford Economic Papers*, 1 (1938), 83-123.
Makower, H., J. Marschak, and H. W. Robinson, 'Studies in the Mobility of Labour: Analysis for Great Britain', Part I, *Oxford Economic Papers*, 2 (1939), 70-97; Part II, *Oxford Economic Papers*, 4 (1940), 39-62.
Marshall, J. D. *Furness and the Industrial Revolution* (Barrow, 1958).
Martin, R. and R. H. Fryer. *Redundancy and Paternalist Capitalism* (London, 1973).
Melling, J. 'Non Commissioned Officers: British employers and their supervisory workers 1880-1920', *Social History*, 5 (1980), 183-222.
Mitchell, G. D., T. Lupton, M. W. Hodges, and C. S. Smith, *Neighbourhood and Community* (Liverpool, 1954).
Mogey, J. N. *Family and Neighbourhood: Two Studies in Oxford* (Oxford, 1956).
Moorhouse, H. F. 'The Marxist Theory of the Labour Aristocracy', *Social History*, 3 (1978), 61-82.
Newby, H. *The Deferential Worker* (London, 1977).
Obelkevich, J. *Religion and Rural Society: South Lindsey 1825-75* (Oxford, 1976).

Orlans, H. *Stevenage* (London, 1952).
Osborn, F. J. *Green Belt Cities* (London, 1946).
— and A. Whittick. *The New Towns: the Answer to Megalopolis* (London, 1969).
Owen, A. D. K. 'The Social Consequences of Industrial Transference', *Sociological Review*, 29 (1937), 331-54.
— 'Rejoinder', *Sociological Review*, 30 (1938), 414-20.
Parkin, F. 'Working class Conservatives: A theory of political deviance', *British Journal of Sociology*, 18 (1967), 278-90.
Parkin, F. *Class, Inequality and Political Order* (London, 1971).
— (ed.). *The Social Analysis of Class Structure* (London, 1974).
Pickett, K. G. and D. K. Boulton. *Migration and Social Adjustment* (Liverpool, 1974).
Polsby, N. W. *Community Power and Political Theory* (New Haven, 1963).
Pollard, S. 'Barrow in Furness and the 7th Duke of Devonshire', *Economic History Review*, 2 s 8 (1955-6), 213-21.
— 'Factory Discipline in the Industrial Revolution', *Economic History Review*, 2 s 16 (1963-4), 254-71.
— 'The Factory Village in the Industrial Revolution', *English Historical Review*, 79 (1964), 513-31.
— *The Genesis of Modern Management* (London, 1965).
— and J. Salt (ed.). *Robert Owen, Prophet of the Poor* (London, 1971).
Porteous, J. D. 'The nature of the company town', *Transactions of the Institute of British Geographers*, 51 (1970), 127-42.
— 'The Company Town of Goole', University of Hull *Occasional Papers in Geography*, 12, 1969.
Reid, A. 'Politics and economics in the formation of the British working class: a reply to H. F. Moorhouse', *Social History*, 3 (1978), 347-61.
Rodwin, L. *The British New Towns Policy* (Cambridge, 1956).
Samuel, R. (ed.). *Miners, Quarrymen and Saltworkers* (London, 1977).
Scase, R. (ed.), *Industrial Society; Class, Cleavage and Control* (London 1977).
Schaffer, F. *The New Town Story* (London, 1970).
Stacey, M. *Tradition and Change, a Study of Banbury* (London, 1960).
— E. Batstone, C. Bell and A. Murcott. *Power, Persistence and Change: a second study of Banbury* (London, 1975).
Stedman Jones, G. 'Class expression versus social control', *History Workshop*, 4 (1977), 163-70.
Sutcliffe, A. *A History of Modern Town Planning* (Birmingham, 1977).
Thomas, B. 'The movement of labour into SE England 1920-32', *Economica*, NS 1 (1934), 220-41.
— 'The influx of labour into London and the South East 1920-36', *Economica*, NS 4 (1937), 323-36.
— 'The influx of labour into the Midlands 1920-37', *Economica*, NS 5 (1938), 410-34.

Thompson, F. M. L. *English Landed Society in the Nineteenth Century* (London, 1963).
Van Onselen, C. *Chibaro* (London, 1976).
Walshaw, R. S. 'The Time Lag in the recent Migration Movements within Great Britain', *Sociological Review*, 30 (1938), 278-87.
Willmott, P. *The Evolution of a Community* (London, 1963).
Young, T. *Becontree and Dagenham* (London, 1934).
Young, M. and P. Willmott. *Family and Kinship in East London* (London 1957).

(*d*) *Royal commissions and reports*

Report of the Select Committee on Coal, PP 1873 x, 1-392.
Census of England and Wales, County of Nottingham, 1921, 1931, 1951.
Royal Commission on the Coal Industry 1925 (Samuel Commission), PP 1926 xiv, 1-310 and Minutes of Evidence (2 vols.).
Royal Commission on the Distribution of the Industrial Population (Barlow Commission), PP 1939-40, 263-592.
Report of the Technical Advisory Committee on Coal (Reid Committee), PP 1944-5 iv, 315-476.
Ministry of Fuel and Power: North Midland Regional Survey Report, HMSO 1945.
New Towns Committee (Reith Committee), Interim Report, PP 1945-6, xiv, 679-700; Second Interim Report, PP 1945-6 xiv, 701-26; Final Report, PP 1945-6 xiv, 727-809.
New Towns Development Corporations Annual Reports, PP *passim*.

Index